A Guide to Wireless Sensor Networks

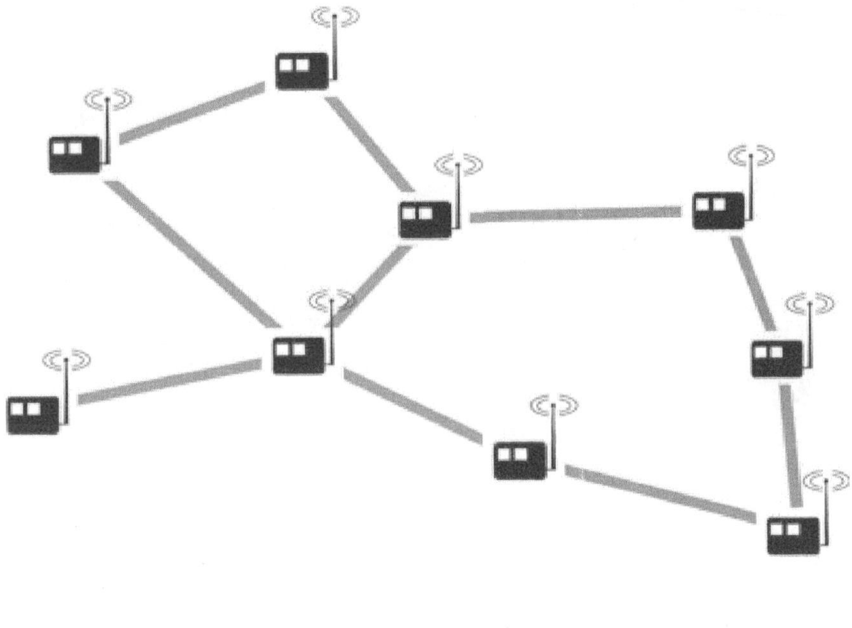

A Guide to Wireless Sensor Networks

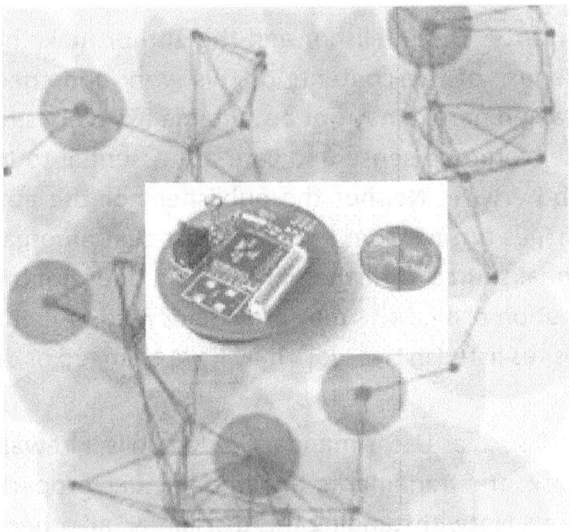

By

S. SWAPNA KUMAR

Professor
Department of Electronics and Communication Engineering
AXIS College of Engineering and Technology,
Thrissur, Kerala.

UNIVERSITY SCIENCE PRESS
(An Imprint of Laxmi Publications Pvt. Ltd.)
An ISO 9001:2015 Company

BENGALURU • CHENNAI • GUWAHATI • HYDERABAD
KOCHI • KOLKATA • LUCKNOW • RANCHI
NEW DELHI

Printed and bound in India

First Edition : 2013; *Edition* : 2025

ISBN **:** 978-93-81159-46-0

Branches

✆	Bengaluru	080-26 75 69 30
✆	Chennai	044-24 34 47 26
✆	Guwahati	0361-254 36 69
✆	Hyderabad	040-27 55 53 83
✆	Kochi	0484-405 13 03
✆	Kolkata	033-40 04 77 79
✆	Lucknow	0522-430 36 13
✆	Ranchi	0651-224 24 64

Published in India by

UNIVERSITY SCIENCE PRESS

(An Imprint of Laxmi Publications Pvt. Ltd.)

An ISO 9001:2015 Company
113, GOLDEN HOUSE, GURUDWARA ROAD, DARYAGANJ,
NEW DELHI - 110002, INDIA

Corporate Office : B-33, Sector-5, Noida,
Gautambuddha Nagar (U.P.) - 201 301, INDIA
Telephone : +91-120-44 46 900
www.laxmipublications.com
info@laxmipublications.com

Dedicated to

My Father Late K. [illegible], mother Late [illegible] Kumar

and to

My Mother [illegible], Wife [illegible]

Children [illegible] and [illegible] Kumar

Dedicated to

My Father Late K. Surendran, brother Late S. Sudarsh Kumar

and to

My Mother K.S.Omana Kutty, Wife Sona S. Kumar,

Children S. Adarsh Roshan and Ardra S. Kumar

CONTENTS

PREFACE

In the recent years, the wireless sensor networks is a fast emerging and growing study areas that have attracted considerable research attention globally. There has been tremendous development in design and development of application related interfaces with sensor networks. Sensor network find applications in several domains such as medical, industrial, military, home networks, space and so on. Wireless sensor networks have moved from the research domain to the real world implementation with wider range of applications.

Wireless Sensor Networks is the fast emerging revolution in wireless communication. Many research people believe that wireless sensor networks can become as important as the Internet in the near future. Just as the Internet allows access to digital information anywhere, wireless sensor networks will provide remote interaction with the physical world. It is becoming possible thanks to the new Wireless Sensor Networks (WSN) will support ubiquitous computing.

Wireless sensor network key changes is the scalability of network protocols, design of energy efficient protocols, design of data handling, localization techniques, sensor nodes design, development of exciting real time new applications that exploit the prospective of wireless sensor networks. This describes how to build these networks, from the layers of the communication protocol through the design of network nodes. This overview summarizes the multiple protocol study, applications of wireless sensor networks, and discusses network device understanding for the successful performance of these applications.

Author realized that with the rapid changing technology and the research development there is need to bring a comprehensive book that can give the guidelines of the concept of wireless sensor network. The potential audience for this book is intended for the college and universities project practicing, graduate students, research scholars and engineers; those who are interested to survey articles on specific topics without searching in larger text books.

The purpose of this book is to present a collection of excellent chapters from leading researchers on various aspects of wireless sensor networks. The book consists of thirteen chapters that provide a comprehensive coverage that is covers the study on wireless sensor network basic concepts. These chapters describe the various characteristics, features and usage of the sensors. It covers the various areas of research study and the survey made related to sensor and protocol fields. Every effort has been made to make the treatment simple and comprehensives. Throughout the book, the value has been given on fundamental concept through illustrative examples and figures.

The writing style of this book is straightforward and makes complex concepts and processes easy to follow and understand. In addition, it offers several features that help readers grasp the material and then apply their knowledge in designing their own wireless sensor network systems. The book is written in a very simple and logical way. Every care has been taken to eliminate language and errors. In spite of all precautions and there might crept some misprint due to oversight. So, the author would appreciative if the reader would bring any such error to their notice. Also, any suggestion for improvement will be thankfully acknowledge.

The author takes these opportunities and is in debt to the large number of books and journals that consulted in the preparation of the book.

The author will like to appreciate the sincere effort and contribution made by various reference authors, scientists who design and developed protocols and contributed various concept related to the wireless sensor network.

Finally, author feels that the reader of this book will gains valuable insight into the main stream ideas behind the technology and instigate them to go forth with innovated new ideas and technologies.

Lastly, but not the least, the author heartily thankful to LAXMI PUBLICATIONS (P) LTD. for taking interest and bringing out the well formatted book in time and pricing it moderately for easy reachable to the readers.

—Author

ACKNOWLEDGEMENT

Authors of this book sincerely thank all the research scholars, scientists, authors, and publishers who worked in the areas of wireless sensor network and the resources have been the source of enlightment, inspiration and guidance in presenting this book. Getting a thoughts and inspiration in writing this text book is from Dr. George Thottappilly and Dr. C.S. Madhu. So, I acknowledge them.

I acknowledge my research guide Dr. M. Nanda Kumar- PhD (IIT Madras), Dr. Sheeba V.S-PhD (NIT, Calicut), and Dr. V. Sumathy- PhD (Anna University), who initially put forth the idea of my research works. This book is result of their initiative and constant motivation.

I received great inspiration and constant encouragement from Dr. Vineeth Paleri (Professor-NIT, Calicut); Dr. Elizabeth Sherly-Director and Asst. Professor, IIITM-K, Techno park; Dr. , Tony Thomas, NTU, Singapore; Dr. K.R. Kashwan, HOD- E and C, Sona CET, Salem; Dr. R.S.D. Wahidabanu, HOD-E and C, Govt. Engg College, Salem; Dr. T. Purushothaman, Govt.Engg College, Coimbatore; Dr. Babu A.V-NIT Calicut; Dr. Sameer S.M-NIT Calicut; Dr. Gylson Thomas-MES Kuttipuram, and Dr. C.N Khairnar, Military College of Telecomm. Engineering, Indore and is highly acknowledged.

I am specially indebt to my teacher Prof. B.S. Kedilaya (HOD- E and C), KVG; faculty members of KVG CET and the Centre for Research, Anna University Coimbatore for providing the platform for where I am today. I have been fortunate in receiving support, assistance and encouragement at various stages from Mr. C.B.K Bhushan- CBK Infotech, Bangalore; Ranjeet Patro-Technical Lead, Samsung Electronics, Bangalore; Mr. Ashish Shukla-Technical Lead, Marvell India Pvt Ltd.; Mr. Sanjeet-NIT; Prof. K.T Joseph-Director Sahrdaya CET; Shri G. Harikumar-Member Secretary, Trust members, Faculty and Staff members of AXIS College of Engineering and Technology for the preparation of the manuscript.

I am indebted to my parents for their immense blessing throughout my life to achieve the success. This book is dedicated for my Late father and Late brother whom I shall forever cherish memories of their love, affection and guidance. Throughout the period while writing this book my mother, wife, son and daughter motivated, supported, and encouraged me with inspiring words.

I also acknowledge Mr. R. Purushothaman, Mrs. P Rajamma, all relatives, and friends for their constant source of encouragement. Special mention is made for all students of AXIS College of Engineering and Technology who gave me support. My thanks are also due to the publishers, editorial and production teams of the LAXMI PUBLICATIONS (P) LTD, for carefully processing the manuscript.

—**Author**

0 INTRODUCTION

0.1 OVERVIEW OF WIRELESS SENSOR NETWORKS

Wireless Sensor Networks (WSNs) have been widely considered as one of the most important technologies for the twenty-first century. WSNs development, pervasiveness and research increased tremendously in the last decade. These are daily used in many different fields, such as precision agriculture, environmental and health monitoring, home automation, military applications, object tracking, intrusion detection and surveillance systems. Such networks are based on embedded devices equipped with sensors and radio transceivers, which communicate and collaborate to achieve a common goal. These embedded devices have many characteristics and constraints such as they are low-power, memory-constrained, low-cost and low-size devices.

Those properties, combined with the low cost and size of these devices and the minimal need of human interaction, led to a great success of embedded systems. However, there are many known drawbacks and constraints like limited resources, error-prone wireless medium, environmental conditions variability, and low security and privacy. The strength of wireless sensor networks lies in the ability to deploy large numbers of tiny nodes that assemble and auto configure themselves.

In the past decade, WSNs have received tremendous attention from both academia and industry all over the world. A large amount of research activities have been carried out to explore and solve various design and application issues, and significant advances have been made in the development and deployment of WSNs. It is envisioned that in the near future WSNs will be widely used in various civilian and military fields, and revolutionize the way we live, work, and interact with the physical world.

0.2 FEATURES OF THIS BOOK

The development of the wireless networking in various areas leads to the most important aspects in the design of WSNs, which involves a variety of network architectural and protocol design issues. The application layer protocols for sensor networks remain a largely unexplored region so far. This is because the focus of the research community is towards the lower layers of the OSI stack rather than towards the higher layers. A new suite of network protocols need to be developed to address the unique characteristics and constraints, in particular, the energy constraint, in sensor networks.

This book focuses on the major basic concepts and wide areas and issues in the design of WSNs, including medium access control, transport protocols, routing and data dissemination, node clustering, node localization, sensor standards, sensor security and harvesting energy.

The aim of this book is to provide a comprehensive and systematic introduction of the fundamental concepts, major issues, and effective solutions in understanding the WSNs from basic level to researchers' motives.

The main features of this book include the following:

- Giving an insight into wireless sensor networks from protocol design perspective.
- Providing a comprehensive and systematic explanation of concepts, major issues, and effective solutions in wireless sensor networking.
- Discussing the methodology and state-of-the-art technologies of various design features in wireless sensor networking.
- Contribution by researchers in the field that is covering the chapters.
- Including a comprehensive up-to-date bibliography.

0.3 ORGANIZATION OF THIS BOOK

This book is organized into 13 chapters that give the comprehensive views of wireless sensor network.

Chapter 1 serves as an introduction to the wireless sensor network (WSN). This chapter includes the basic concept; Ad-Hoc network characteristics, typical network architecture, and IEEE technological background are introduced.

Chapter 2 presents a brief overview of sensors, different types of sensor architectures and introduces a wireless sensor topology and its applications.

Chapters 3 discuss on different layer protocol and, clustering and its related feature. Also discussed, the optimization of layer protocols for WSNs is presented. The major network design objectives and challenges are described.

Chapter 4 is dedicated to medium access control (MAC) in WSNs. The fundamental overview and concepts on MAC protocols for wireless networks are introduced; the major challenges in energy management on MAC design for sensor networks, proposed MAC protocol are discussed.

Chapter 5 focuses on fundamental concepts on routing and the major challenges in routing are discussed. Moreover, a survey of routing design challenges protocols for WSNs is presented.

Chapter 6 is dedicated to data aggregation in WSNs. The concepts of the importance of data aggregation are discussed. The chapter also presents an overview of design areas on data aggregation techniques for WSNs.

Chapter 7 is dedicated to energy efficiency and power control in WSNs. The concepts of power management, the major challenges in WSNs are introduced. The different model on sensor is discussed.

Chapter 8 focuses on node localization in WSNs. The importance of node localization is introduced, the major challenges in node localization are discussed, and an overview of typical localization algorithms is presented.

Chapter 9 presents an overview of standardization activities and relevant standards for WSNs, focused on the IEEE 802.15.4, ZigBee and IEEE 1451.5 standards.

Chapter 10 concentrates on test-bed in WSNs. The purpose of test-bed fundamental concepts for WSNs is presented for implementation and analysis.

Chapter 11 is dedicated to network security in WSNs. The importance of security in WSNs, the major challenges in designing security mechanisms are discussed, and variety of effective security techniques for WSNs are presented.

Chapter 12 addresses energy harvesting in WSNs. The importance of energy harvesting is explained, and introduced effective energy harvesting for WSNs are presented.

Chapter 13 presents an overview of the application of the wireless sensor network paradigm, as well as future aspect of WSNs.

1

INTRODUCTION TO WSN

1.1 INTRODUCTION

The advancement of science and technology deeply intertwined the growth of communication era. In the recent year the revolution of personal computers, mobile telephony and the Internet, changed the face of wireless communication world. Such field of wireless networking integrates into the areas of personal computing, cellular technology, and the Internet. This is due to the increasing demand and interactions between communication and computing. Moreover, the computing is to a level of a high quality and speed. This change the face of the information technology and its access with the logo "anytime anywhere" into "all the time, everywhere."

Wireless sensor network is a small spatially distributed network devices that can communicate with each other over the wireless medium. Wireless sensor networks development and pervasiveness gave origin to a wide range of different applications with different features and needs. On the other hand, research today mainly focuses on development, optimization and improvement of physical, MAC and routing layer issues, parameters adjustment, in order to minimize the energy consumption and maximize the lifetime, scalability and security of the systems rather than on implementing or designing an application support

1.2 WIRED AND WIRELESS NETWORKS

There are two different types of networks that are available today are Wired and Wireless networks. Wired are differentiated from wireless as being wired from point to point.

1.2.1 Wired Networks

In the wired network generally connected with the help of wires and cables. It usually established with the help of physical devices like Switches and Hubs in between to increase the strength of the connection. Generally, the cables being used in this type of networks are CAT5 or CAT6 cables. These networks are usually more efficient, less expensive and much faster than wireless networks. Once, the connection is set there is a very little chance of getting disconnected.

Advantages of Wired Networks are:

- A wired network offer connection speeds of 100Mbps to 1000Mbps
- Physical, fixed wired connections are not prone to interference and fluctuations in available bandwidth that can affect some wireless networking connections.

Disadvantages over Wireless Networks are:

- Maintenance and cost of the network cables between computer systems is high, whenever there any failure in the cables occurs for the replacement.
- Wired network will limit the logical reason for the purchasing of a laptop in the first place.

1.2.2 Wireless Networks

Wireless networks use some sort of radio frequencies in air to transmit and receive data instead of using some physical cables. The most admiring fact in these networks is that it eliminates the need for laying out expensive cables and maintenance costs.

Advantages of Wireless Networks are:

- Mobile users are provided with access to the network real-time information even when they are away from their homes or offices.
- Setting up a wireless system is easy and fast and it eliminates the need for pulling out the cables through walls and ceilings.
- Network can be extended to places, where the accessibility of wired not present.
- Wireless networks provide more flexibility and easily adapt to changes in the configuration of the network.

Disadvantages of Wireless Networks are:

- Interference due to weather, other radio frequency devices, or obstructions like walls.
- When multiple connections exist, then the total Throughput is affected.

1.3 PROBLEMS IN WIRELESS COMMUNICATIONS

In a multipath Propagation, a signal travels from its source to destination, in between there are obstacles make the signal propagate in paths beyond the direct line of sight due to reflection, refraction, diffraction and scattering. Path loss is the attenuation of the transmitted signal strength as it propagates away from the sender, and it is determined as the ratio between the powers of the transmitted signal to the receiver signal.

Some of the problems related to wireless communication are multipath propagation, path loss, interference, and limited frequency spectrum. It is sometimes important to estimate the path loss in wireless communication networks. Due to the radio frequency and the nature of the terrain are not same everywhere, it is hard to estimate the path loss during communication. During communication a number of signals in the atmosphere may interfere with each other resulting in the destruction of the original signal.

1.4 WIRELSS COMMUNICATION REQUIREMENTS

Communication in the wireless link, need to follow up the packet-based wireless communication protocol as illustrate in the Fig. 1.1. A transmitter must carefully modulate the RF carrier while receiver performs demodulation and signal analysis. It is important to note that many of the operations must be performed in parallel with each other, where the distinct layers overlap in time.

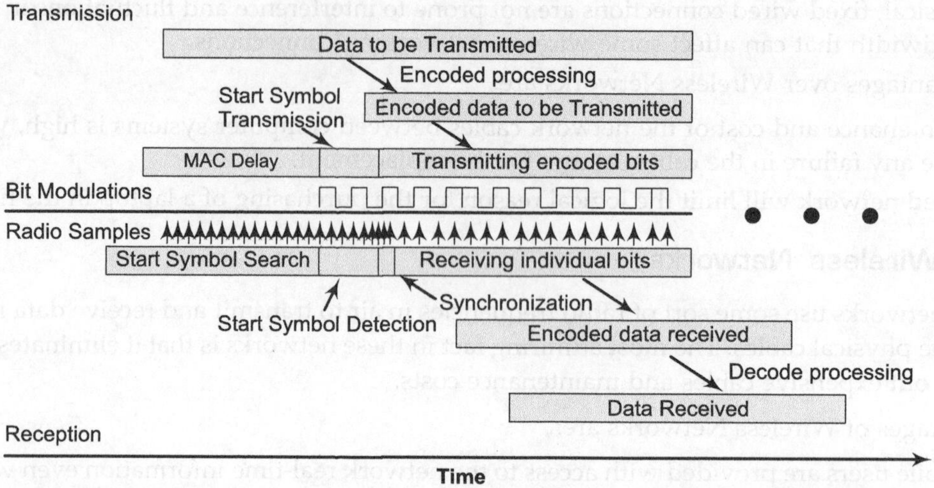

Fig. 1.1 Phases of wireless communication for transmission and reception

The first step in the communication process is to encode the data for transmission. The coding schemes are designed to increase the probability of a successful transmission by preventing and correcting slight errors. In either, a collection of one or more data bits, called a symbol, are coded into a collection of radio transmission bits called chips by Manchester encoding.

The media access control (MAC) protocols are designed to allow transmission, where multiple transmitters to share a single communication channel. One of the simplest MAC protocols is carrier sense media access (CSMA), where each transmitter first checks for an idle channel prior to each transmission. If the channel is busy, it waits for a short, random, delay after which it reinitiates the transmission.

The first piece of data packet to be actually transmitted to the receiver over the radio link is synchronization symbol or start symbol signals to the receiver to determine the timing of the arriving transmission. The start symbol is immediately followed by the encoded data transmitted as a serial stream. As the transmission proceeds, the transmitter must precisely control the timing of each bit transition so that the receiver can maintain synchronization. Skewed bit transitions can cause the sender and receiver to get out of synch, resulting in an unsuccessful transmission or corrupted data.

For a receiver, the first part of data reception is to detect the transmission. In order to properly detect the start symbol, the receiver to filter out background noise and must sample the channel at least twice the radio chip rate. Otherwise, the relative phase of the sampling and the transmission may result in the receiver missing the start symbol.

Once detected, the receiver must then synchronize itself to the exact phase of the incoming transmission allowing the receiver to determine the start and end of each bit window used by the transmitter. Synchronization requires the incoming transmission to be sampled higher than twice the bit rate so the timing of the bit transitions can be determined. Once synchronized, the receiver then samples the value of the incoming signal at the center of each bit. Finally, the individual bits are extracted from the radio, are assembled into blocks are decoded back into the original data and assembled into a packet are the encoded version of actual data messages.

1.5 TYPES OF WIRELESS NETWORK

Besides hardware technologies, the development of WSNs also relies on wireless networking technologies. Wireless network is a network set up to communicate among computers and other network devices by using radio signal frequency. It is an IEEE 802.11 protocol standard also referred to as Wi-Fi network or WLAN. The 802.11 protocol, standard for wireless local area networks (WLANs), was introduced in 1997. It was upgraded to 802.11b with an increased data rate and CSMA/CA mechanisms for medium access control (MAC). Although, designed for wireless LANs that usually consist of laptops and PDAs, the 802.11 protocols are also assumed by many early efforts on WSNs. However, the high power consumption and excessively high data rate of 802.11 protocols are not suitable for WSNs.

Recently, the 802.15.4-based ZigBee protocol was released, which was specifically designed for short range and low data rate wireless personal area networks (WPAN). Its applicability to WSNs was soon supported by several commercial sensor node products, including MicaZ, Telos, and Ember products.

A wireless network is basically a set of nodes that form a connected network via wireless communication. Each node has a transmitter and can reach all nodes in its transmission area, which we call direct communication. The two main components are wireless router or access point and wireless clients.

There are several types of wireless networks developed in the last few years, only two major types of wireless networks exist:

1. Cellular Networks and
2. Ad-Hoc Networks.

At present, a large variety of networks exists, ranging from the well-known infrastructure of cellular networks to non-infrastructure wireless ad-hoc networks. Here, describe these types of networks and their corresponding tasks in the following.

1.5.1 Cellular Networks

A cellular network is an asymmetric radio network made up of fixed transceivers (or nodes) which maintain the signal while the mobile transceiver which is using the network is in the vicinity of the node. Cellular Networks have stationary base stations with a certain transmission range that divide the communication field into cells. Each node belongs to a specific cell and therefore to the corresponding base station. If a node is located in an overlap area, it generally belongs to the base station with the strongest signal.

As displayed in Fig. 1.2 a base station is only responsible for the nodes that are in its cell got exchange of information's and control actions. The communication between nodes in the same cell goes via the base stations, using a direct connection from the source to the base station and from the base station to the destination. To establish communication between nodes in different cells, first the source transmits communication data directly to the base station which in turn communicates to the base station to which the destination node belongs. Finally, the base station of the destination communicates directly to the sink node.

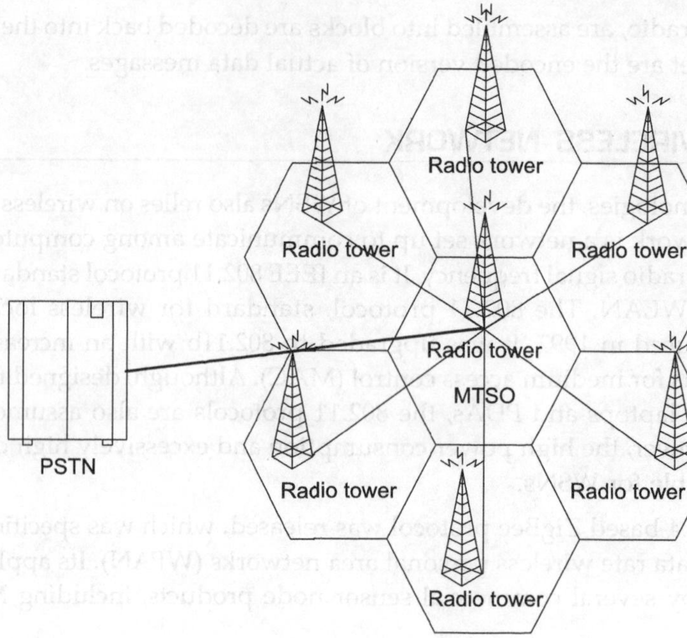

Fig. 1.2 Cellular network

The most important task of a cellular network is long distance communication between mobile devices, like mobile telephones.

1.5.2 AD-HOC Networks

Ad-Hoc networks are wireless, self-organizing systems created by certain nodes defined within communication range to form a temporary networks. Nodes topology is dynamic, decentralized, ever changing and may move around arbitrarily. Ad-hoc networks, also called mesh networks, are organized to provide pathways for data to be routed from source system to and from the desired destination.

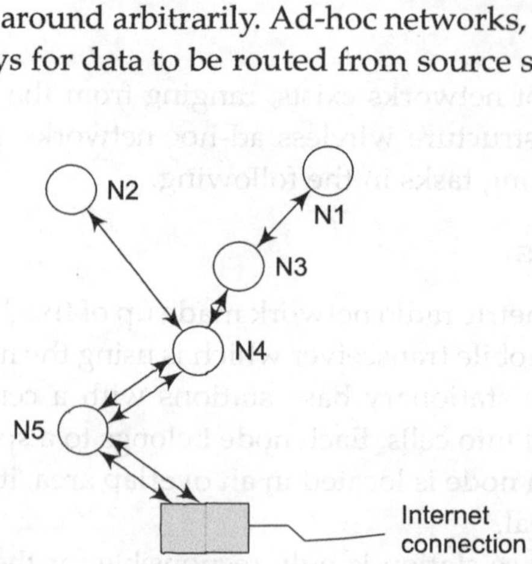

Fig. 1.3 Basic structure of an ad-hoc network

The ad-hoc network connection in the wireless mode shows the path that makes the peer-to-peer connectivity as shown in Fig. 1.3. Here, there is no access or base point that means no central configuration of network. These networks can be connected directly or to through intermediate control or monitoring network when the intermediate point breaks, then the network is automatically reconfigure to form the alternate path. Typically, all available nodes represent as N_1, N_2 ... are network users that share the data transfer.

1.6 AD-HOC NETWORKS TOPOLOGY

Ad-Hoc Networks are wireless networks in which the determination of which nodes have to forward data is dynamically based on the network topology as displayed in Fig. 1.4. An ad-hoc infrastructure network does not have access points for passing information between participants through a central information hub that can be a hardware device or software on a computer.

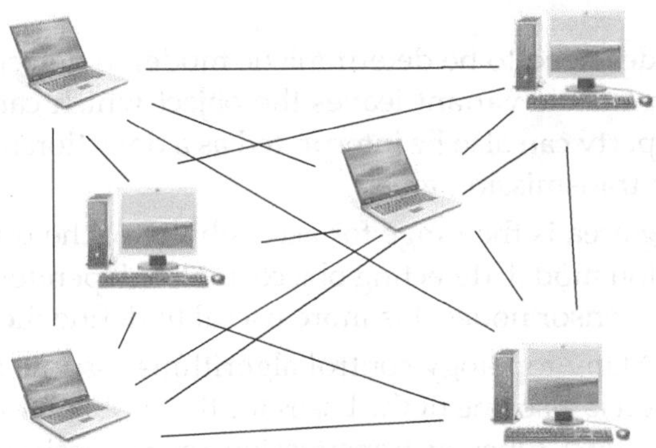

Fig. 1.4 Ad-hoc networks

An ad-hoc network is "a transitory association of mobile nodes which do not depend upon any fixed support infrastructure. Here, as per the above Fig. 1.4 the mobile laptop is not dependent on any fixed nodes whereas the laptop when keep on moving across the range called as dynamic movement then also the transfer of data from the source to sink takes place.

Three features must be present in an application for it to deserve the ad-hoc scenario are:

1. Mobility: This allow the device user to use the application everywhere, the user should not be limited by range. The range limit is set by the business logic of the application.
2. Peer-to-Peer: It's a direct communication between peers which is mandatory in the ad-hoc mode. This means that the client/server relationship is defined in an ad-hoc manner by the application logic.
3. Collocation: All logical interactions between ad-hoc applications result in a physical interaction between users the service has to be location-based.

There is several research study carried out for the protocol development in the ad-hoc networking. Ad-hoc and sensor network provides an opportunity for the researchers in the field of electronics, computer and mathematics particularly to disseminate and carry out research in this rapid emerging field.

1.6.1 Ad-Hoc Network Properties

The energy constraint is the basic factor of the wireless network. As wireless nodes work on battery source, they have limited energy supply. The network is needed to operate for a longer period to increase the life span of battery and termed as 'Long Network Lifetime'. Some of the important properties of ad-hoc network are as follows:

1. Network Structure: The network structure can be Flat or Hierarchical. In a flat structure each node has the same functionality and role. In a hierarchical structure (Ex. in cellular network)some nodes are designated special tasks, like collecting information or running an algorithm to detect the neighborhood structure.
2. Sensor Deployment: There are many ways to deploy sensor nodes, varying from randomly to exact locations. This determines the sensor nodes density is high or normal, that distinguish working sensor from redundant sensors. If the total number of nodes is on the same order as the number of working sensors it is called a normal sensor density; if the total number of sensors is orders of magnitude higher than the working sensors it is called a high sensor density.
3. Detection Model: Nodes need to be deterministic model to detect every object in its sensing range. The non-deterministic variant leaves the object which cannot be detected. For non-sensor nodes this property can also be interpreted as a detection model for communication if a neighbor falls in its transmission area.
4. Sensing Area: Sensing area is the range for accessibility of the transmitter to the receiver in addition to the detection model, detecting objects is also dependent of the sensing area of the sensor nodes. For non-sensor nodes it is more useful to define the type of transmission area.
5. Transmission Range: Many topology control algorithms assume an adjustable transmission power. This assumption is also one of the bases for the problem to maximize the lifetime of a network. The three distinct types of transmission ranges assignment are: continuous, discrete and on-off.
6. Time Synchronization: Time synchronization is then necessary in order to define time limited rounds and time stamps to topology control and it is mostly used for waking up from sleep mode, and other state to control algorithms.
7. Failure Model: In the real world a node failure due to interruption runs out the battery energy. For example, such destruction is not unlikely to happen when the wireless network is used in the military field. The assumption with respect to node failure is important to manage the topology to regain back.
8. Sensor Mobility: The sensor nodes have certain degree of mobility for each network, varying from low-mobile to high-mobile. There exist several mobility models for the sensor nodes.
9. Location Information: The specific location information, *i.e.*, the geographical location, can be used to determine the overlapping of sensing areas and to compute the distance between neighbors. In stationary wireless networks, where the nodes are placed at selected locations, the location information can be easily coded.

Mobile situations in randomly distributed nodes need equipment like GPS or specific location algorithms to detect their positions.

1.6.2 Advantages of an Ad-Hoc Networks

- Independent: Ad-hoc network is independence from central network administration, self-configuring, nodes acts as routers, self-healing and re-configures itself.
- Accessibility: Ad-hoc network provides access to information and services regardless of geographic position.
- Deployment: The networks can be set up at any place and time.
- Flexible: Ad-hoc network being able to access the Internet from many different locations

- Infrastructure-less: The networks work without any pre-existing infrastructure. This allows people and devices to interwork in areas with no supporting infrastructure.
- Scalable: Ad-hoc network accommodates the addition of more nodes, hence it is scalable.
- Dynamic: Ad-hoc network can freely and dynamically self-organize into arbitrary and temporary network topologies.

1.6.3 Limitation of Ad-Hoc Networks

Thou ad-hoc networks are having greatest importance on its advantages, there are still some limitations:

- Ad-Hoc Networks reliability requires a sufficient number of available nodes. Sparse networks can have problems.
- Each node must have peer-to-peer connectivity for full performance, then it will be effective.
- Ad-Hoc Networks throughput is affected by system loading.
- Large networks can have excessive latency (time delay), which affects some applications.

1.7 AD-HOC NETWORK DESIGN ISSUE AND CHALLENGES

Ad-hoc wireless networks inherit the traditional problems of wireless communications, such as bandwidth optimization, power control, and transmission quality enhancement, while, in addition, their mobility, multi-hop nature, and the lack of fixed infrastructure create a number of complexities and design constraints that are new to mobile ad hoc networks. Certain issues and challenges are listed below:

(*i*) Infrastructure less network: The ad-hoc wireless network basic fundamental is its lack of infrastructure; also, lack of centralized mechanism brings added difficulty in fault detection and correction.

(*ii*) Dynamic Topology: The ad-hoc wireless network is dynamically changing nature of mobile nodes causes to the formation of an unpredicted topology. This topology change causes frequent route change, network partitioning and packet dropping.

(*iii*) Limited Link Bandwidth and Quality: The ad-hoc wireless network is significantly lower capacity leads mobile nodes communicate each other via bandwidth-constrained, variable capacity, error-prone, and insecure wireless channels, causes more problematic network congestion.

(*iv*) Energy Constrained Operation: Energy constraints are another big challenge in ad-hoc wireless network design. These constraints in wireless network arise due to battery powered nodes which cannot be recharged. This becomes a bigger issue in mobile ad-hoc networks because as each node is acting as both an end system and a router at the same time, additional energy is required to forward packets.

(*v*) Robustness and Reliability: In MANET, network connectivity is obtained by routing and forwarding among multiple nodes. Due to various conditions when a node fails it misbehaves nodes and unreliable links impact on overall network performance. Moreover, this increases the design complexity significantly.

(*vi*) Network Security: Mobile wireless networks are more vulnerable to information and physical security threats than fixed-wired networks. In addition, because a mobile ad-hoc network is a distributed infrastructure-less network, the security control is hard to implement.

(*vii*) Quality of Service: Quality of Service (QoS) guarantee is very much essential for the successful communication of nodes in the network. The dynamically changing topology, limited bandwidth and quality makes difficulty in achieving the desired QoS guarantee for the network on throughput, packet loss, delay, and jitter and error rate.

1.8 MANET

An ad-hoc network is a collection of wireless mobile hosts forming a temporary network without the aid of any stand-alone infrastructure or centralized administration. Mobile Ad-hoc networks (MANE) are self-organizing and self-configuring multihop wireless networks that changes dynamically. This is mainly due to the random mobility of the nodes in the network to access wireless channel. The node in the network not only acts as hosts but also as routers that route data to/from other nodes in network in multihop way.

Mobile ad-hoc networks open the window for the fixed devices to establish networks on the fly, *i.e.*, formally, a MANET is a collection of mobile devices which form a communication network with no pre-existing wiring or infrastructure.

The key element of a MANET is that it is a self-organizing structure that allows nodes to join or leave the network, resulting in a continuously changing network topology. They allow the applications running on these wireless devices to share data of different types and characteristics. As shown in Fig. 1.5 the topology of the MANET is distributed and Ad-hoc operation.

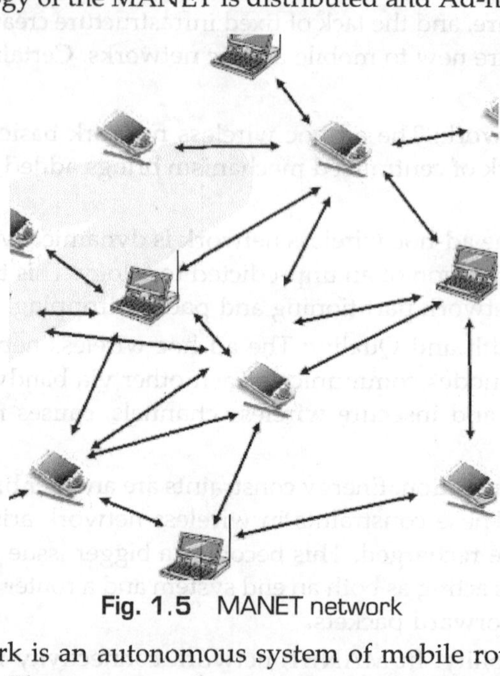

Fig. 1.5 MANET network

A mobile Ad-hoc network is an autonomous system of mobile routers (and associated hosts) connected by wireless links. The routers are free to move randomly and unpredictably to organize themselves depending upon the network's wireless topology that changes rapidly and arbitrarily. MANETs use a multihop approach for communication between two non-neighboring nodes in the network.

In mobile ad-hoc networks, where there is no infrastructure support as is the case with wireless networks, and since a destination node might be out of range of a source node transmitting packets;

a routing procedure is always needed to find a path so as to forward the packets appropriately between the source and the destination. The base station in a cell, without routing can reach all mobile nodes via broadcast in common wireless networks. In the case of ad-hoc networks, each node must be able to communicate data by forwarding it to other nodes. This creates additional problems of dynamic topology which is unpredictable connectivity changes.

MANETs are fundamental structures that can be used for other applications to build upon. The nodes in a MANET are having energy constraint due to battery and therefore it has a limited transmission range. Mobile ad-hoc network is an infrastructure-less multi-hop network where each node communicates with other nodes directly or indirectly through intermediate nodes. Fig. 1.6 shows an example mobile ad-hoc network and its communication topology.

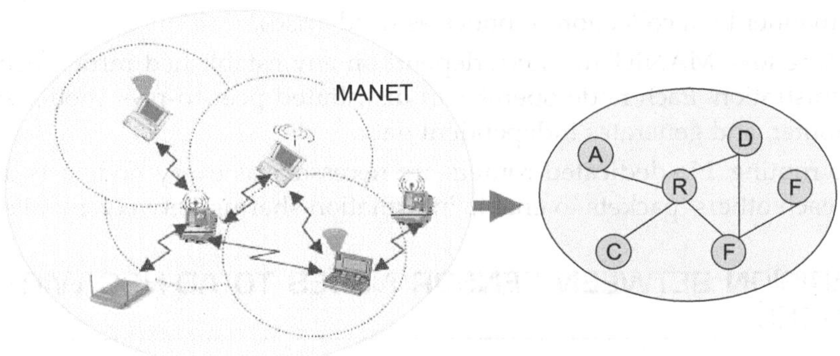

Fig. 1.6 Mobile ad-hoc networks

All nodes in a MANET basically function as mobile routers participating in some routing protocol required for deciding and maintaining the routes. In general, ad-hoc network may have multiple hops routes between nodes, hence, such networks is called "multi-hop wireless ad-hoc networks.

Here, nodes A,C,R,D,E and F are mobile routers.

1.8.1 Constraints in MANET

In ad-hoc networks, depending on routing tables changes in topology and routing algorithms need to be adapted. This is difficult due to the bandwidth constraint and dynamic topology. The support of Quality of Service (QoS) in MANETs is a challenging task. Following are the some of the constraints in such network.

- Asymmetric links: Most of the wired networks relay on the symmetric links that is always fixed. But, in the case of ad-hoc networks, the nodes are mobile and constantly changing their positions within network.
- Routing Overhead: In wireless Ad-hoc networks, nodes often change their locations dynamically within the networks. So, some stale routes are generated in the routing table node, which leads to unnecessary routing overhead.
- Interference: This is the one of the major problem with mobile ad-hoc networks as links come and go depending on the transmission characteristics, one transmission might and can corrupt the total transmission.
- Dynamic Topology: This is also the major problem, where ad-hoc routing topology is not constant due to change in medium characteristics and the dynamic mobile node change.

1.8.2 Characteristics of MANET

MANET is having the general characteristics of wireless network and additional characteristics that are specific to the Ad-Hoc Networking. Some of the characteristics are as follow:

- Mobility: Each node is free to move about while communicating with other nodes. The topology of such network is ad-hoc and it is dynamic in nature. Due to constant movement of the participating nodes, there is continuous change in the intercommunication patterns among nodes.
- Wireless: MANET nodes communicate wirelessly and share the same media (radio, infrared, etc.).
- Ad-hoc-based: A mobile ad-hoc network is a temporary network formed dynamically in an arbitrary manner by a collection of nodes as need arises.
- Infrastructure-less: MANET does not depend on any established infrastructure or centralized administration. Each node operates in distributed peer-to-peer mode, acts as an independent router, and generates independent data.
- Multi-hop routing: No dedicated routers are necessary as every node acts as a router and forwards each others' packets to enable information sharing between mobile hosts.

1.9 COMPARISION BETWEEN SENSOR NODES TO AD-HOC WIRELESS NETWORKS

Wireless sensor networks share similarities (and differences) with ad-hoc wireless networks. The main similarity is the multi-hop communication method. The differences among the two types of networks are listed below:

- More nodes are deployed in a sensor network, up to hundred or thousand nodes, than in an ad-hoc network that usually involves far fewer nodes.
- Compared to ad-hoc, sensor nodes are more constrained in computational, energy and storage resources.
- Neighboring sensor nodes often sense the similar events from their environments forwarding the same redundant data to the base station.
- Aggregation and in-network processing often work together between sensor nodes that are not in the case of ad-hoc networks.
- Sensor nodes can be deployed in human intervention free environments and can remain unattended after deployment for a long time.

1.10 APPLICATIONS OF AD-HOC NETWORK

Ad-hoc lead networks of computer/PDA users have become common. Blue tooth compatible system is increasingly drastic. Control systems and industrial process monitoring and control are becoming major applications for mesh networking.

Environments monitoring and information collection are difficult to serve with dedicated wiring, being spread over a large area, often with difficult access. An example of network is shown in Fig. 1.7 depicts a precision agriculture deployment of an active area for application research. Numbers of nodes scattered throughout a field form network to assemble together, establish a routing topology, and transmit data back to a collection point.

Fig. 1.7 Deployment of ad-hoc wireless embedded network for precision agriculture

The application demands for robust, scalable, low-cost and easy to deploy networks are perfectly met by a wireless sensor network to continue and to deliver data.

Sensor networks from small-scale (e.g. household security monitoring) to large scale (*e.g.* wildlife tracking) are also being developed with ad-hoc networking as the operational structure. This shows that ad-hoc networks are the most efficient way to maintain system-wide communications.

1.11 WIRELESS SENSOR NETWORKS (WSN)

The miniaturization of electronics, along with the advances in wireless communications and the development of multi-functional sensors, has lead to the birth of a new technology named Wireless Sensor Networks (WSNs). Many people confuse WSNs with the ad-hoc network. In the reality, WSNs are unlike Ad-hoc networks in the sense that WSNs are resource limited, they are deployed densely, they are prone to failures, the number of nodes in WSNs is several orders higher than that of ad-hoc networks, WSN network topology is constantly changing, WSNs use a broadcast communication mediums and finally sensor nodes don't have a global identification tags.

The present electronics era is now migrating from the personal computer to the ubiquitous computing age, where the wireless network provides easier solution for the interconnection. This gives the rapid growth in the wireless technologies. One such area of ad-hoc network and other is Wireless Sensor Networks. The short distance to long distance computing in wireless mote is the greatest role of Ad-hoc and WSN.

Figure 1.8 shows the Wireless Sensor Nodes and Networks that composed of a large number of small nodes with sensing, computation, and wireless communication capabilities. These sensor nodes are small in sizes, but are equipped with sensors, embedded microprocessors, and radio transceivers. It have not only sensing capability, but also data processing and communicating capabilities. Sensor nodes are usually scattered and the sensor network protocols and algorithms must provide self-organizing capabilities to find network position. The coordination of sensor nodes produces high-quality information about the sensing environment.

Fig. 1.8 Wireless sensor nodes and networks

A wireless sensor node consists of a microcontroller, a radio, several sensors, storage and a battery. A WSN is composed of sensor nodes that sense several environmental phenomena and form an ad-hoc network for the purpose of collaboratively processing and transmitting the data to the interested parties. A WSN is a self-organizing network that does not need user intervention for configuration or setting up routing paths. Therefore, WSNs can be used in virtually any environment, even in inhospitable terrain or where the physical placement is difficult.

WSN technologies are embedding sensing, processing and communication in one tiny device. It is an array of small, locally battery powered sensor nodes that wirelessly communicate information sampled from surrounding environment to a access point in the form of sink or gateway and via a wireless communication channel. Whereas, the Ad-hoc network form the network of peer-to-peer connection of computer without access point. Information arrives at the base station is then transmitted outside the network to be processed further for the purpose of satisfying the objective of WSN operation.

In the Fig. 1.9 represents the WSN architecture, where the user of different network communicated with its destination target via an internet to the wireless sensor nodes in a defined path.

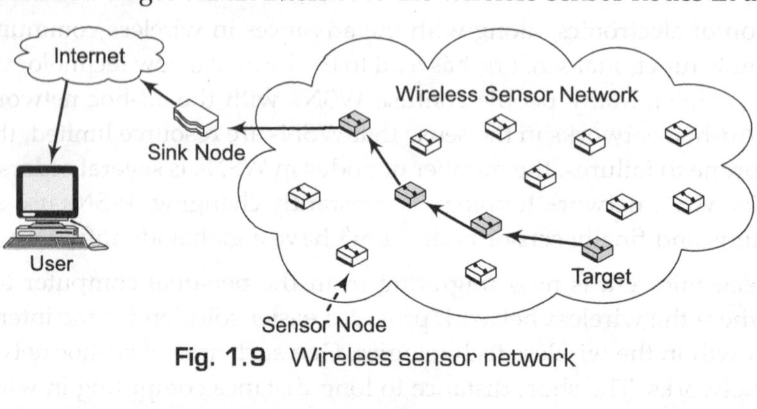

Fig. 1.9 Wireless sensor network

1.12 INTERNAL ARCHITECTURE OF WSN

In WSNs, each wireless sensor node called mote in the network is equipped with a sensor module that enables to capture signals generated by the events through a sensor channel, and a network to send information to the base station through a wireless channel. The Internal architecture of a wireless sensor node is shown in Fig. 1.10.

A WSN typically comprises a large number of spatially distributed, tiny, embedded sensor devices that are networked to cooperatively collect, process, and deliver data about a phenomenon that is of interest to the users.

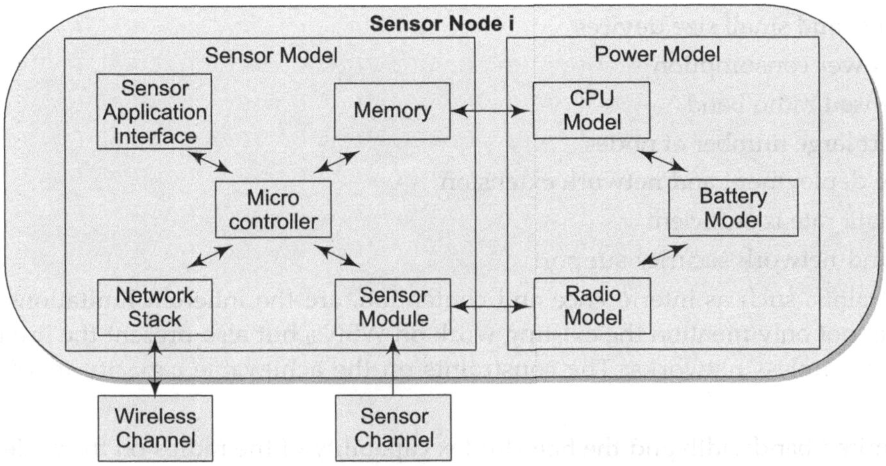

Fig. 1.10 Internal architecture of a wireless sensor node

The combination of the sensor module, the network stack with a radio frequency transceiver having its operational states of, idle and sleep, the sensor application interface, the processing or microcontroller module, and multiple types of memories (program, data, or flash memories for storage capability) is usually called the sensor function model.

1.13 WSN VS. TRADITIONAL WIRELESS NETWORKS

There are many existing protocol, techniques and concepts from traditional wireless network, such as cellular network, mobile ad-hoc network, wireless local area network and Bluetooth, are still used in wireless sensor network, but there are also many fundamental differences that lead to the need of new protocols and techniques. Some of the most important characteristic differences are summarized below:

- Number of nodes in wireless sensor network is much higher than any traditional wireless network. Possibly a sensor network has to scale number of nodes to thousands. Moreover, a sensor network might need to extend the monitored area and has to increase number of nodes from time to time. This needs a highly scalable solution to ensure sensor network operations without any problem.
- Due to large number of sensor nodes, addresses are not assigned to the sensor nodes. Sensor networks are not address-centric; instead they are data- centric network. Operations in sensor networks are centered on data instead of individual sensor node. As a result sensor nodes require collaborative efforts.
- Sensor nodes mainly use a broadcast communication paradigm, whereas most ad-hoc networks are on point-to-point communications.
- Sensor nodes are less expensive than nodes in ad-hoc networks.

1.14 WSN TECHNOLOGY REQUIREMENTS

According to its concept and target applications the wireless sensor network technology should meet the following requirements:

- Low cost and small size devices
- Low power consumption
- Unlicensed radio band
- Support large number of nodes
- Simple deployment and network extension
- Low data rate is sufficient
- Data and network security support

The constraints, such as interference and contention, are the inherent limitations in wireless networks that not only mention the existing work on WSNs, but also present the important solutions in other wireless networks. The constraints on the achievable capacity in WSNs are the following:

- The limited bandwidth and the half-duplex capability of the radios on the nodes.
- Interference and contention on the wireless medium.
- The topology of the network.

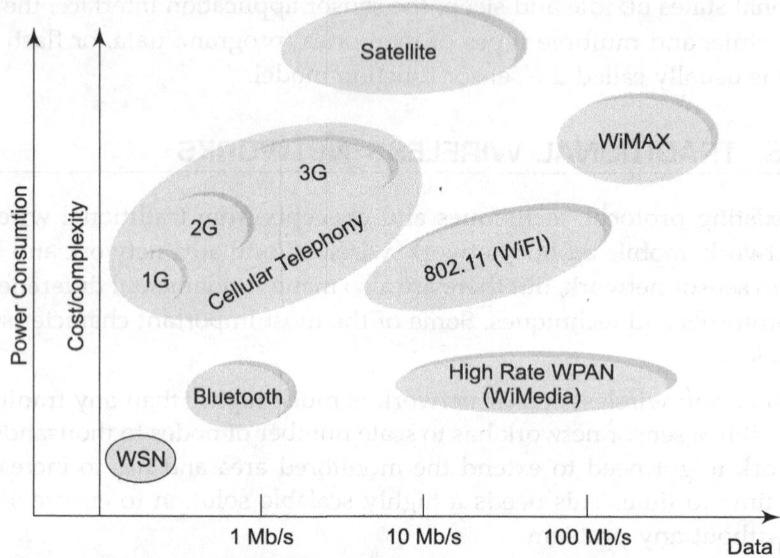

Fig. 1.11 Position of WSN among other wireless network technologies

As shown in the Fig. 1.11 chart the position of WSN among other wireless network technologies shows the low data rates and low power consumption communication device. The protocol of wireless is IEEE 802.11 standards.

To achieve the advantages described above it is sufficient to standardize communication protocols to be used. Such approach is most common in network standards. The most promising and popular standards for wireless sensor networks is IEEE 802.11 specified by Institute of Electrical and Electronics Engineers (IEEE) shown in Fig. 1.12.

Fig. 1.12 Logo of Institute of Electrical and Electronics Engineers (IEEE)

1.14.1 Wireless Sensor Network Characteristics

A WSN typically consists of a large number of low-cost, low-power, and multi-functional sensor nodes that are deployed in a region of interest. Compared with traditional wireless communication networks, cellular systems and MANET, sensor networks have the following unique characteristics and constraints:

- Dense Node Deployment: Sensor nodes are usually densely deployed in a field of interest. The number of sensor nodes in a sensor network can be several orders of magnitude higher than that in a MANET.
- Battery-Powered: Sensor Nodes are usually powered by battery. In most situations, they are deployed in a harsh or hostile environment, where it is very difficult or even impossible to change or recharge the batteries.
- Energy Storage Constraints and Computation: Sensor nodes are highly limited in energy, computation, and storage capacities.
- Self-Configurable: Sensor nodes are usually randomly deployed without careful planning and engineering. Once deployed, sensor nodes have to autonomously configure themselves into a communication network.
- Data Redundancy: In most sensor network applications, sensor nodes are densely deployed in a region of interest and collaborate to accomplish a common sensing task. Thus, the data sensed by multiple sensor nodes typically have a certain level of correlation or redundancy.
- Unreliable Sensor Nodes: Sensor nodes are usually deployed in harsh or hostile environments and operate without attendance. They are prone to physical damages or failures.
- Frequent Topology Change: Network topology changes frequently due to node failure, damage, addition, energy depletion, or channel fading.
- No Global Identification: Due to the large number of sensor nodes, it is usually not possible to build a global addressing scheme for a sensor network, because it would introduce a high overhead for the identification maintenance.
- Many-to-One Traffic Pattern: In most sensor network applications, the data sensed by sensor nodes flow from multiple source sensor nodes to a particular sink, exhibiting a many-to-one traffic pattern.
- Application Specific: Sensor networks are application specific. A network is usually designed and deployed for a specific application. The design requirements of a network change with its application.

The unique characteristics and constraints present many new challenges in the design of sensor networks.

1.15 MULTI-HOP WIRELESS SENSOR NETWORK MODEL

Wireless Sensor Network (WSN) is a low power and small size sensor nodes. These sensor nodes consists of spatially distributed autonomous can sense, process and communicate information among them. Fig. 1.13 represents Typical Multi-hop Wireless Sensor Network Architecture. The sensors cooperatively monitor physical or environmental conditions, such as temperature, sound, vibration, pressure, motion or pollutants.

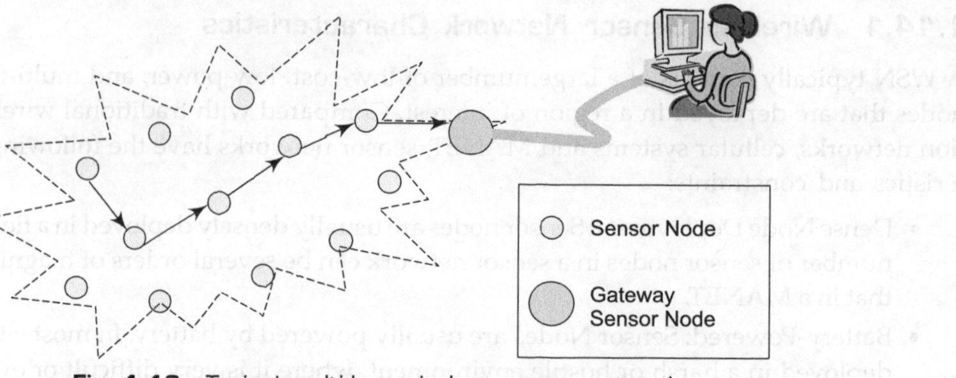

Fig. 1.13 Typical multi-hop wireless sensor network

A sensor network is a WSN each sensor supports a multi-hop routing algorithm to forwarders, relaying data packets to a base station. Each node in a sensor network is typically equipped with a radio transceiver other wireless communications device, a small microcontroller, and an energy source, usually a battery.

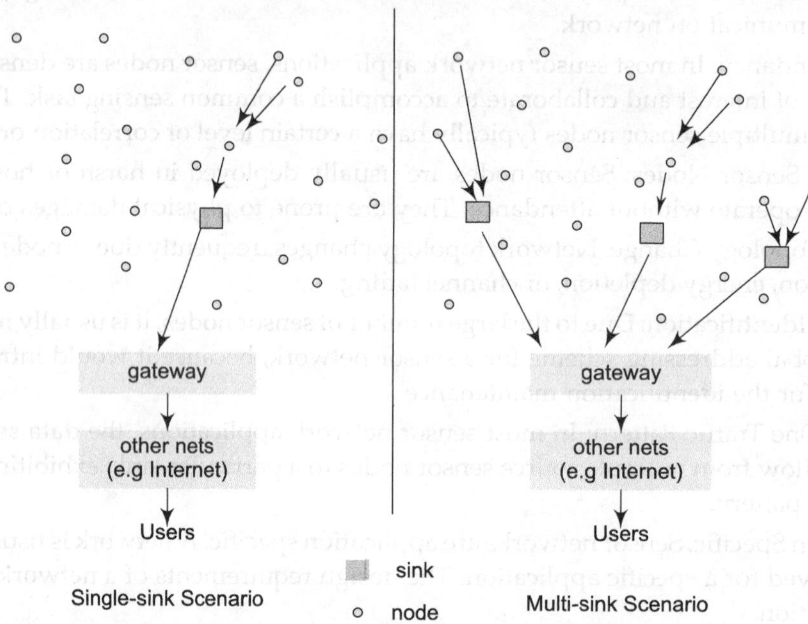

Fig. 1.14 Left part: single-sink WSN. Right part: multi-sink scenario

In a WSN nodes sense the environment and communicate through the wireless links. The data is forwarded, via multiple hops, to a sink locally or is connected to other networks (*e.g.*, the Internet) through a gateway. The nodes can be stationary or moving and can be aware of their locations or can be homogeneous or not.

Fig. 1.14 shows the traditional single/multi-sink WSN. Given a level of node density, multiple-sink WSN can be scalable than a single-sink network. In many cases, nodes send the data collected to one of the sinks that forward the data to the gateway, toward the final user. Therefore, the multiple sinks ensure better network performance with respect to the single-sink case, but the communication protocols must be more complex.

1.15.1 Sensor Network Units

The sensor network is different from Ad-hoc networks. It is usually composed of nodes of several orders of magnitude higher than Ad-hoc networks. These nodes consist of four basic units embedded on to them is categorized as follow:

- Sensing units: This is the primary units of sensor nodes as it fulfills the objectives of observing the target or environments. There may be one or more sensors that track the parameters of the pressure, temperature etc.
- Processing unit: The micro-controller is responsible for performing the controlling functions such as activating sensors, analog to digital signal conversion, temporary data storage memory, raw data aggregation, enabling and disabling of transceiver, generating clock cycle for control functions.
- Transceiver unit: This unit is for communication which comprises of combined circuitry of a transmitter and a receiver. The nodes don't transmit or receive the data packet simultaneously as it function as half duplex. However, the transceiver units consume energy during transmitting and receiving.
- Power units: This is the life of sensor nodes, without which the purpose of sensor nodes is baseless. It consists of battery unit as a source for the sensor node.

The silent features of the sensor nodes are application dependent. It is deployed for monitor or sensing the signal from the harsh environment. The nodes are densely deployed to protect the network from the failure of nodes. Data from the nodes can be aggregated and collected by the special nodes called the base station or the sink.

1.16 HARDWARE OF MOTE

The sensor nodes are the spatially distributed sensor node are called Golem Dust or Mica2Dot nodes as shown in Fig. 1.15, use small batteries as energy storage. Such sensors are agreeably to monitor physical or environmental conditions, such as temperature, motion, sound, pressure etc. A typical WSN device consists of the following components:

(a) Golem Dust (b) Mica2Dot

Fig. 1.15 Wireless sensor motes

- Sensor/Actuator Boards: Some sensor nodes are Chipcon CC1000 and CC2420, and Nordic NrF905. Typical examples of sensors are temperature, light, humidity, acoustic, pressure, chemical sensors and accelerometers. Examples of actuators can be speakers, LEDs, buzzers.
- The Wireless Transceiver available on sensor nodes is usually a low-rate, low-power, short-range radio. The transceivers mostly operate on unlicensed bands like the 868- 915MHz or

2.4 GHz industrial, scientific and medical (ISM) bands. Typical data rates supported by the radios are 50-250 Kbits per second.

- The Processor used on the sensor nodes is required for processing the sensed data, running the system software and the networking protocol stack. Mostly, 8-bit or 16-bit processors run on specialized operating systems such as AmbientRT, TinyOS and Contiki.
- Memory/Storage capabilities are also quite limited. Usually, a few Kbytes of RAM and a few tens of Kbytes of flash RAM are available for storing data and code.
- The Power Supply of the sensor nodes generally consists of batteries. In many WSN applications it is impractical or impossible to replace/recharge the batteries of the nodes so energy harvesting methods available on the nodes.

1.16.1 Hardware Structure of a Wireless Sensor Node

A wireless sensor network application needs the specialized hardware that has an embedded hardware structure. In the structural diagram as shown in Fig. 1.16, it shows the wireless sensor node.

A controlling element of a wireless sensor is the Micro Controller Unit (MCU), where its requirements are for short wake-up time, low power consumption, high computing speed. The MCU includes Flash and SRAM memory. An embedded operating system and other software are written into the MCU flash memory. The MCU controls the connected transceiver and provides arbitration for analog and digital sensors. The other required interfaces are, such as USB, SPI, and RS232.

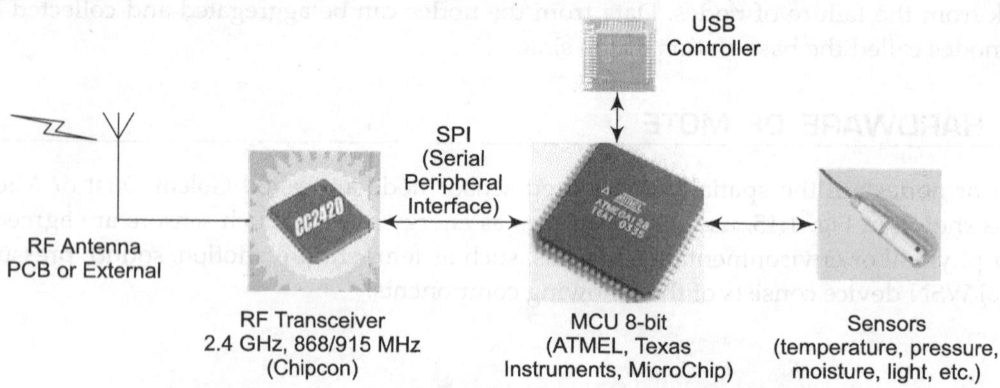

Fig. 1.16 Hardware structure of a wireless sensor node

The most popular micro controllers are produced by Texas Instruments, Intel, Atmel, Microchip, and Philips. The next important part is the RF transceiver (including the antenna) and it is controlled by the MCU. RF transceivers, produced by Chipcon, are widely used in the development and evaluation.

Low power consumption is one of the main requirements for transceivers, because the largest amount of energy is spent for radio transmission and reception which is met with the help of routing protocols. A routing is used for efficient power consumption.

1.17 WSN SYSTEM ARCHITECTURE

A wireless ad hoc network is an autonomous system consisting of mobile hosts connected by wireless links. Such networks can be quickly and widely deployed to serve a multiplicity of purposes. OSI layer is the basic ground rule that is essential when developing algorithms for computer communication networks, *i.e.*, networking protocols. The algorithm for the layer of the OSI (Open Systems Interconnection) model ensures that every layer operates completely independently. Thus, the data flow diagram in Fig. 1.17 illustrates how the various components in our sensor network architecture exchange information with each other.

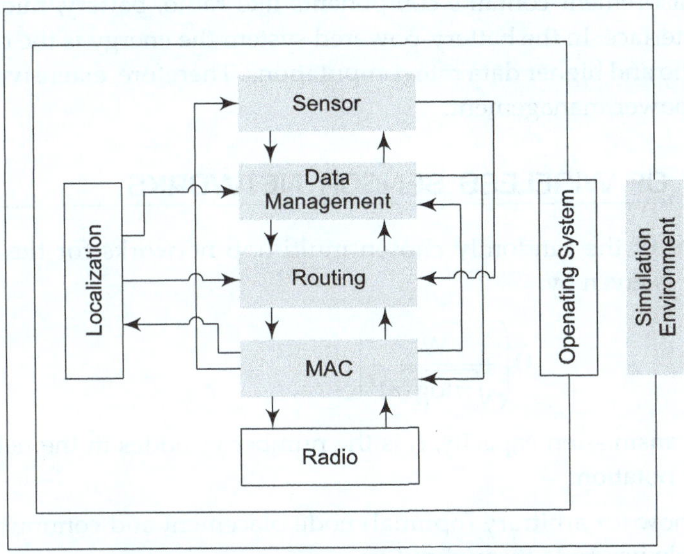

Fig. 1.17 Architecture of a wireless sensor node

It is immediately apparent that the level of inter-dependence between the various components is much greater than the layered approach used in conventional computer communication networks. For example, the localization component may estimate the location of a node by using connectivity information provided by the MAC layer. This location estimate may then be used by the routing component to perform geographic routing. The total number of required transmissions to send one data packet from a source to a sink must be minimized.

In the Fig. 1.18 shows a common WSN architecture, the measurement nodes are deployed to acquire measurement such as temperature, voltage, etc., the nodes that is the part of a wireless network

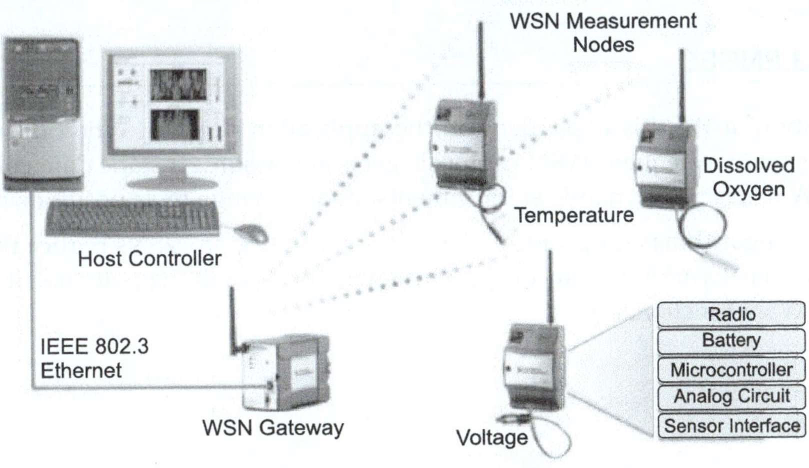

Fig. 1.18 Common wireless sensor network architecture

administered by the gateway that governs network aspect such as client authentication and data security. The collected data from each node are sends over a wired connection, typically Ethernet, to a host controller. There, software such as the NI Lab VIEW graphical development environment can perform advanced processing and analysis and present that data to meet the end user.

Often a WSN measurement contains components like radio, battery, microcontroller, analog circuit, and sensor interface. In the battery-powered system the energy is the constraint factor due to frequent use of radio and higher data rate computations. Therefore, extensive research evolves in the areas of battery power management.

1.18 CAPACITY OF WIRELESS SENSOR NETWORKS

The capacity of WSN of the randomly chosen multi-hop networks for the destinations nodes throughput per node is given by:

$$\Theta\left(\frac{W}{\sqrt{n\log(n)}}\right) \qquad ...(1.1)$$

Where, W is the transmission capacity, n is the number of nodes in the network and Θ represents the asymptotic notation.

Also, the results show for arbitrary (optimal) node placement and communication patterns. In this case the achievable per node throughput is:

$$\Theta\left(\frac{W}{\sqrt{n}}\right) \qquad ...(1.2)$$

The capacity can be improved by the mobility of the nodes which can reduce the number of hops between the source and the destination and in turn reduce the contention in the network. Further improvements on the obtained capacity bounds by introducing relay nodes which do not generate traffic but act as routers to deliver data to the destination.

When we switch to WSNs, the traffic is usually results in many-to-one communications. These affects the reuse possibilities of the medium and the schedule that the nodes are transmitting with.

1.19 CHALLENGES

The functionality of a WSN is dependent on the application domain. When the changes in the network topology are infrequent WSN protocols generally assume a static data collection pattern and when the WSNs is in a dynamic environments, nodes change their positions in real-time.

Consequently, algorithms and protocols for self-organization in WSNs reduce the negative impact of mobility on networking protocols performance, and exploit the potential of using mobility to enhance the WSN functionality.

1.19.1 Challenges of Sensor Networks

The nodes are subjected to the following limitations depending upon the applications.

- Energy constraints: Wireless sensors are operational subjects to the amount of battery source. The battery source consumes most of the energy source for the sensing, processing, and transceiver units. However, most of the energy is consumed by the transceiver unit for communicating data. Recently, energy harvesting in sensor nodes attracted a lot of interest. So, it is imperative for the nodes to conserve energy to improve the system life.
- Processing limitation: The processing power of sensor nodes is directly linked to the amount of energy available in sensor networks. The challenges are constraint in the computation power usage, the processing CPU units' energy consumption than the transmission unit of the node. So, to control of energy wastage, the computation power should be limited based on optimal algorithm design under real-time situation. .
- Communication limitation: The nodes are establishing the communication between transceivers using RF links. Due to the limitation of the nodes computation capabilities the data gets corrupted when simultaneous transmission from nodes takes place. Further, the bandwidth is limited for certain applications. So, the nodes take the short multi-hop communication between nodes instead of long range communication to conserve energy.

  The four major sources of energy waste in sensor communication are highlighted as follow:
  - Collision: When the simultaneous transmission and receiving of packets takes place by the nodes, there are chances of corrupted packet, so the transmitting node need to notify and send the packet again that cause the wastage of energy.
  - Overhearing: The node that receives the packet of other nodes receive packet that are not addressed to them. This will cause the nodes processor to be in active which result in energy wastage.
  - Control packet overhead: When the sensor node communicated with respect to large packet size with respect to short control packet, it consumes energy for transmission, acknowledging the receipt of packet and permission for packet sending.
  - Idle listening: When nodes are not transmitting or receiving packets, the sensor node radio module also consume energy in its idle mode. Studies have observed that in idle mode, a radio module consume 50-10% energy required for receiving.

1.19.2 Challenges of WSN

The challenge is to achieve these objectives given the distinguishing characteristics of WSNs, including:

- Physical Constraints: The nodes are often highly constrained in their communicational, computational, energy, and storage resources, and hence there is need to manage in most efficient manner.
- Dynamism: The WSN environment is highly dynamic and the network topology can sometimes vary during operation since nodes may fail, be moved, or be added at any time.
- Deployment in Hostile Environments: The communication links between dynamics nodes are subject to additional noise and interference. The dynamism-related problems further aggravated within hostile and inaccessible environments as the nodes are more likely to fail.
- Scalability: WSN nodes are often relatively small and scalable since it is deployed in large numbers and at high densities to produce high data rates and fidelity.

The ideal wireless sensor network is scalable, consumes very little power, and is smart and software programmable. It is capable of fast data acquisition, reliable and accurate over the long term, costs little to purchase and install, and requires no real maintenance.

1.19.3 Challenges of Ubiquitous Computing

Further in the following the major challenges of the WSNs face in order to contribute to the ubiquitous computing vision are:

- Heterogeneity: With the collaboration among sensor nodes with different hardware capabilities offers more flexibility and supports, but at the same time forces algorithms and protocols to become resource aware. Therefore, resources in a WSN have to be discovered and efficiently managed for an improved network functionality and performance.
- Dynamics: The paradigm shift from static sensor arrays to pervasive applications involves a major increase in the overall degree of mobility or dynamics. Mobility has a negative effect on the quality of wireless communication and the performance of networking protocols, for the enhancement of network performance.
- Proactive and transparency: Proactive WSNs have the potential of delivering context-aware and just-in-time services. The challenge is to provide the user with information and services in a transparent manner. WSN-based proactive services should ideally assist the user in an unobtrusive way and at the same time ensure efficient utilization of resources.

In addition the traditional WSN challenges of sensor nodes is to assure a long network lifetime and scalability to improve network performance.

1.20 APPLICATIONS

The range of WSNs applications has extended considerably, mainly due to the following reasons:

- The processing capabilities of the nodes have evolved up to a point that enables them to execute complex tasks and to make decisions autonomously;
- A group of sensor nodes can combine their resources and capabilities through collaboration and provide complex services, such as reliable event detection, localization or tracking; and
- An interoperable collection of heterogeneous devices can achieve superior functionality by using the flexibility of the resource-lean devices in conjunction with the enhanced capabilities of the more endowed nodes.

Some of the specific areas, where the data collection from complex environment of WSNs is as shown:

- Environmental monitoring: Environmental monitoring is the traditional WSN application, where a static array of sensors is randomly or uniformly deployed over an area to gather sensor readings and to transmit them at a central point for processing. Typical settings include precision agriculture, habitat monitoring or ocean water monitoring.
- Animal monitoring: WSNs nodes are attached to animals for the purpose of studying their behavior, location or confine them within an area. Examples include wild life monitoring and cattle herding.
- Health care: Sensor nodes integrated into garments, also known as Body Area Networks (BANs), as shown in Fig. 1.19 can be used to monitor the vital signs of patients, their walking pattern, or even to locate the patients or medical personnel inside a building.

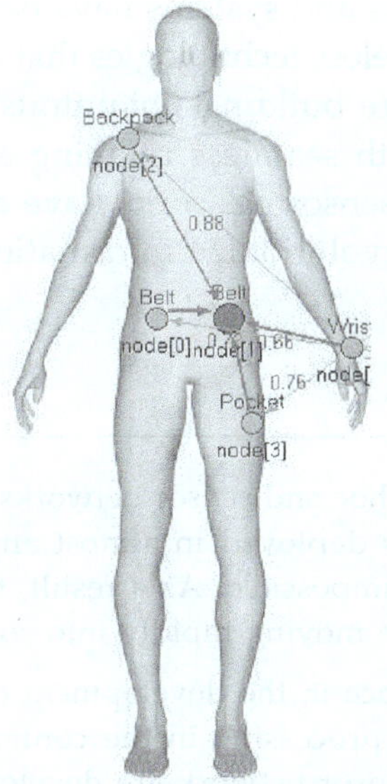

Fig. 1.19 Body Area Networks

- Industrial safety: Industrial safety can benefit from the WSN technology. Sensor nodes can collaboratively determine and prevent potential hazardous situations, and alert or take action at the point of interest. For example, in the oil and gas industry, dangerous situations may arise by storing incompatible substances in close proximity of each other or exceeding the maximum storage volume threshold for hazardous substances.
- Smart buildings: Sensor networks can provide monitoring and control of environmental conditions in buildings (such as temperature, humidity, or light), electronic door and way signs localization of people.
- Emergency: Emergency applications have as main objective to rescue of people in danger. For this purpose, people at risk carry a sensor node that permits localization in case of disaster. For example, rescue of avalanche victims and fire fighting and rescue.
- Military: WSN represents a promising technology for military applications with a sensor networks are: battlefield surveillance, mapping opposing terrain, nuclear, biological and chemical attack detection and reconnaissance, target tracking.
- Transport and logistics: Transport and logistics represent an important market for WSNs. The goal is to monitor the storage conditions of products, to verify the loads, and to real-time localize the goods at production sites, distribution centers or stores.

The development of wireless sensor networks was motivated by military applications such as battlefield surveillance and is now used in many industrial and civilian application areas, including industrial process monitoring and control, machine health monitoring, environment and habitat monitoring, healthcare applications, home automation, and traffic control.

The design, implementation, and operation of a sensor network basically depend upon inputs, networking and protocols, embedded control systems, and algorithms. Such networks are often deployed in resource-constrained environments, such as battery operated nodes. These constraints dictate that sensor network problems are best approached in a hostile manner, by jointly considering the physical, networking, and application layers and making major design tradeoffs across the layers.

Wireless communications networks and systems have been penetrating in our everyday lives, spurring growing interest in new wireless technologies that are smart for tomorrow. Next generation 4G wireless network systems are building ubiquitous mobile infrastructure and integrate multimedia services applications with seamless roaming across networks. Based on computational capacity, wireless ad-hoc and sensor networks have attracted more and more attention in recent years. These networks will revolutionize information gathering and processing in both rural and urban environments.

SUMMARY

The focus of this book is on wireless ad-hoc and sensor networks. Wireless sensor networks have already entered many aspects of our lives. It is deployed in almost any environment, especially those where conventional wired sensor systems are impossible. As a result, the last few years the research areas on ad-hoc and wireless sensor networks are moving rapidly into commercialization and standardization.

There is extensive research taking place in the development of new algorithms for data aggregation, ad-hoc routing, and distributed signal processing in the context of wireless sensor networks. As the algorithms and protocols for wireless sensor network are developed, they must be supported by a low-power, efficient and flexible hardware platform.

Twenty years from now, wireless sensors will be a "behind-the-scenes" technology that has grown to impact every aspect of our lives. All factory and machine command and control systems will have switched over to wireless sensing and control points. The building control and automation systems will be replaced by an invisible wireless mesh. The devices will be fully automated and self configured. Tomorrow's world will be wireless sensor network technology and the systems that will grow to impact every aspect of our global changes.

QUESTIONS

1. Distinguish the difference between wired, wireless and wireless sensor networks.
2. What are the phases of wireless communication during transmission and reception?
3. What is the difference between wireless and ad-hoc network?
4. State the challenges of wireless sensor network and also ad-hoc networks.
5. Draw and explain the architecture of wireless sensor networks.
6. Explain the hardware structure of wireless sensor nodes.
7. What are the challenges of WSN?
8. Briefly, explain the applications of wireless sensor networks.

BIBLIOGRAPHY

- Roland Flury and Roger Wattenhofer, "Routing, Anycast, and Multicast for Mesh and Sensor Networks" INFOCOM 2007.
- J. Heidemann, F. Silva, C. Intanagonwiwat, R. Govindan, D. Es- trinand D. Ganesan, "Building ecient wireless sensor networks with lowlevelnaming," in Proceedings of the eighteenth ACM Symposium onOperating Systems Principles (SOSP01), pp. 146-159, 2001.
- Wireless Sensor Networks: Principles and Applications Chris Townsend, Steven Arms MicroStrain, Inc.

- Lewis, F.L., "Wireless Sensor Networks," Smart Environments: Technologies, Protocols, and Applications, ed. D.J. Cook and S.K. Das, John Wiley, New York, 2004.
- Tiwari, A., Lewis, F.L., Shuzhi S-G.; "Design & Implementation of Wireless Sensor Network for Machine Condition Based Maintenance," Int'l Conf. Control, Automation, Robotics, And Vision (ICARV), Kunming, China, 6-9 Dec. 2004.
- Lu, B. Blum, T. Abdelzaher, J. Stankovic, and T. He, RAP: A Real-Time Communication Architecture for Large-Scale Wireless Sensor Networks, IEEE Real-Time Applications Symposium, June 2002.
- "Operating Systems for Wireless Sensor Networks: A Survey Technical Report" Adi Mallikarjuna Reddy V AVU Phani Kumar, D Janakiram, and G Ashok Kumar, May 3, 2007.
- L. Girod, J. Elson, A. Cerpa, T. Stathopoulos, N. Ramanathan, and D. Estrin. Emstar: a software environment for developing and deploying wireless sensor networks. In Proceedings of the 2004 USENIX Technical Conference, Boston, MA, 2004.
- Jallad and T. Vladimirova. Operating systems for wireless sensor networks in space. In Proceedings of the 8th Military and Aerospace Applications of Programmable Logic Devices and Technologies International Conference (MAPLD'2003), pages P{1005, Washington DC, US, NASA, September 7-9 2005.
- M. S. 'A survey of applications of wireless sensors and wireless sensor networks' in Intelligent Control, 2005. Proceedings of the 2005 IEEE International Symposium on, Mediterrean Conference on Control and Automation, pages 719{ 724, 2005.
- Venkateswarlu R and Janakiram D. A simple model for evaluating the scalability in wireless sensor networks. In International Conference on Intelligent Sensors, Sensor Networks and Information Processing Conference,, December 2005.
- H. Park, W. Liao, K. H. Tam, M. B. Srivastava, and L. He. A uni_ed network and node level simulation framework for wireless sensor networks. Technical report, September 7 2003.
- M. Beigl, A. Krohn, T. Riedel, T. Zimmer, C. Decker, M. Isomura The uPart Experience: Building a Wireless Sensor Network, IEEE/ACM Conference on Information Processing in Sensor Networks (IPSN), 2006
- S. Wielens, M. Galetzka and P. Schneider, Design support for Wireless Sensor Networks Based on the IEEE 802.15.4 Standard, Personal, Indoor and Mobile Radio Communications (PIMRC), 15-18 Sept. 2008.
- M. Isomura, T. Riedel, C. Decker, M. Beigl, H. Horiuchi, Sharing sensor networks, Sixth International Workshop on Smart Appliances and Wearable Computing (IWSAWC) 2006, Lisbon, Portugal: Proceedings of the ICDCS 2006, IEEE Computer Society.

Basics on Wireless Sensor Networks

2.1 INTRODUCTION

A WSN can be generally described as a network of nodes that cooperatively sense and may control the environment enabling interaction between persons or computers and the surrounding environment. The concept of wireless sensor networks is based on a simple equation:

Sensing + CPU + Radio = Thousands of potential applications

A general definition of a sensor is "a device that produces measurable response to a change in a physical or chemical condition" such as heat, light, or pressure, and generates a signal that can be measured or interpreted". The Sensor Network defines a sensor node, capable of responding to one or several stimuli, processing the data and transmitting the information over a short distance using a radio link. It is an infrastructure comprised of sensing (measuring), computing, and communication elements that gives an administrator the ability to instrument, observe, and react to events and phenomena in a specified environment. Sensor nodes employ electronic circuits that have minimum power consumption. Some of the modern sensors nodes are shown in the Fig. 2.1.

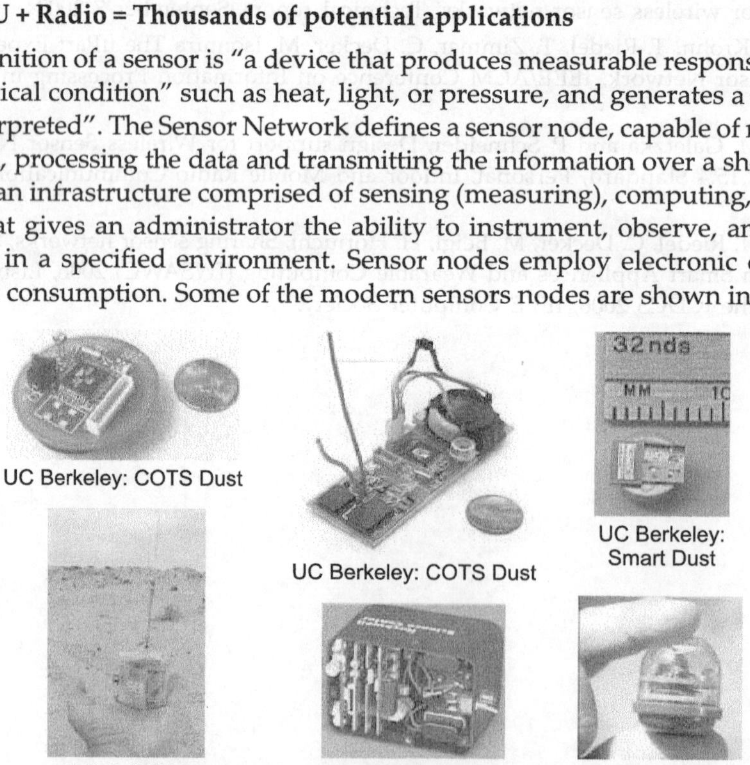

Fig. 2.1 Modern sensor nodes

A Sensor Network is a wireless, ad-hoc network, made of a large number (hundreds or thousands) of nodes, whose positions occur randomly. The OSI model and the classic layered view of communication networks may or may not apply directly to sensor networks. Other models of sensor network communications include a protocol stack model that includes physical, medium access control, network, transport and application layers as well as power management, mobility management and task management planes. Typically sensors are thought of as measuring light, sound and temperature. However, sensors can measure other variables, such as electromagnetic fields or vibrations.

2.2 BACKGROUND OF SENSOR NETWORK TECHNOLOGY

Sensors in a WSN have a variety of purposes, functions, and capabilities. WSNs have shown a remarkable scope in an "exciting emerging domain of low-power wireless motes with a tiny amount of CPU and memory, and large networks for high-resolution sensing of the environment". Sensor networking is a multidisciplinary area that involves, radio and networking, signals processing, artificial intelligence, database management, systems architectures administration, resource optimization, power management algorithms, and hardware and software.

Today's sensors are low-cost low-power multifunctional nodes described as "smart" inexpensive devices equipped with multiple onboard sensing elements. Sensor devices, or wireless nodes (WNs), are called motes. Mote is a development from Berkley's lab.

A WSN consists of densely distributed nodes with high-density that support sensing, signal processing, embedded computing, and connectivity. Sensors span several orders of magnitude in physical size; they range from nanoscopic-scale 1 to 100 nm in diameter devices to mesoscopic-scale 100 and 10,000 nm devices at one end, and from microscopic-scale 10 to 1000 μm devices to macroscopic-scale devices at the other end. The latest generation of sensors, especially the miniaturized sensors that are directly embedded in some physical infrastructures, as micro sensors.

Sensors can be simple point elements or multipoint detection arrays. It may be passive and/or be self-powered; it may require relatively low power from a battery or line feed. Sensors facilitate the controlling of factories, offices, homes, vehicles, cities, and the ambiance, especially as commercial off-the-shelf technology becomes available. The sensor network technology applications such as buildings can "self-detect" structural faults (*e.g.*, fatigue-induced cracks). Earthquake-oriented sensors in buildings can locate potential survivors and can help assess structural damage; tsunami-alerting sensors are useful for nations with extensive coastlines. Sensors also find extensive applicability on the battlefield for reconnaissance and surveillance.

2.3 WIRELESS SENSOR TOPOLOGY

There are a number of different topologies for radio communications networks. A brief discussion of the network topologies that apply to wireless sensor networks are outlined below.

I. **Star Network (Single Point-to-Multipoint):** A star network as shown in Fig. 2.2 is a communications topology where a single base station can send or receive a message from a number of remote nodes. The remote nodes send or receive a message from the single base station; and not permitted to send messages to each other. The advantage of this type of network for wireless sensor networks is in its simplicity and the ability to keep the remote node's power consumption to a minimum. It also allows for low latency communications between the remote node and the base station.

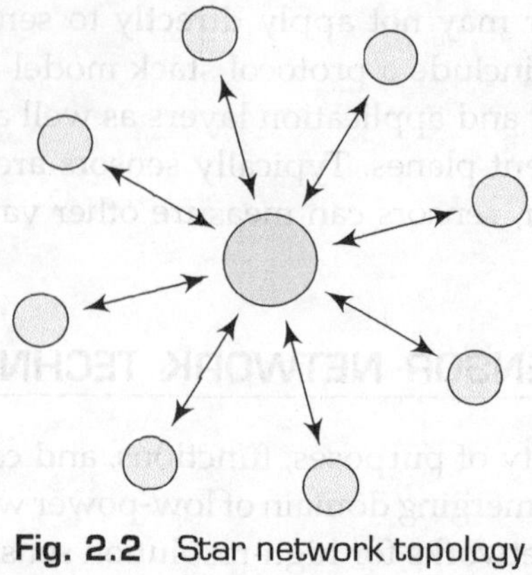

Fig. 2.2 Star network topology

The disadvantage of such a network is that the base station must be within radio transmission range of all the individual nodes and is not as robust as other networks due to its dependency on a single node to manage the network.

II. **Mesh Network:** A mesh network is a pear-to-pear connection that allows any node in the network to transmit to any other node in the network within its radio transmission range. This network topology as shown in Fig. 2.3 has the advantage of redundancy and scalability. Every node communicate with other nodes in its proximity and if an individual node fails, a remote node still can communicate to any other node, which in turn, can forward the message to the desired location.

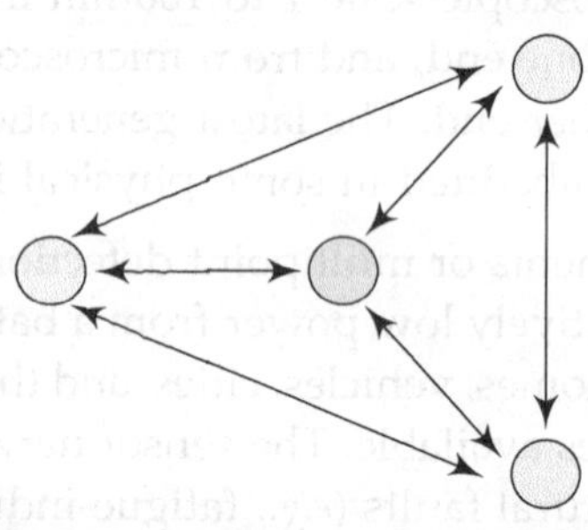

Fig. 2.3 Mesh network topology

The disadvantage of this type of network is in power consumption for the nodes that implement the multihop communications are generally higher than for the nodes that don't have this capability, often limiting the battery life. Additionally, as the number of communication hops to a destination increases, the time to deliver the message also increases, especially if low power operation of the nodes is a requirement.

III. **Hybrid Star - Mesh Network:** A hybrid lies between the star and mesh network. It provides for a robust and versatile communications network, while maintaining the ability to keep the wireless sensor nodes power consumption to a minimum. In the Fig. 2.4 the network topology, the lowest power sensor nodes are not enabled with the ability to forward messages. This allows minimal power consumption to be maintained while forwarding messages from the low power nodes to other nodes on the network. Generally, the topology implemented by the up and coming mesh networking standard known as ZigBee.

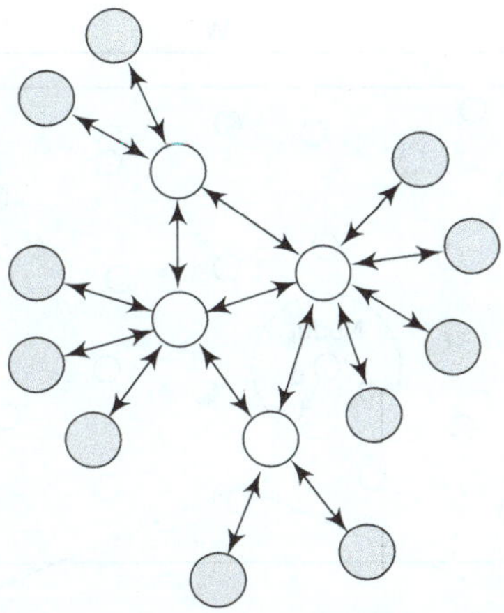

Fig. 2.4 Hybrid star-mesh network topology

2.4 BASICS ON SENSOR

There are four basic components in a sensor network:

I. An assembly of distributed or localized sensors;

II. An interconnecting network (usually, but not always, wireless-based);

III. A central point of information clustering; and

IV. A computing resources to handle data correlation, event trending, status querying, and data mining.

The computation and communication infrastructure associated with sensor networks is often specific to this environment and rooted in the device-and application-based nature of these networks. Following is a sample classification of research topics by frequency of publication the areas covered in WSN articles area in the following area:

Deployment 9.70% Target tracking 7.27% Localization 6.06% Data gathering 6.06% Routing and aggregation 5.76% Security 5.76% MAC protocols 4.85% Querying and databases 4.24% Time synchronization 3.64% Applications 3.33% Robust routing 3.33% Lifetime optimization 3.33% Hardware 2.73% Transport layer 2.73% Distributed algorithms 2.73% Resource-aware routing 2.42% Storage 2.42% Middleware and task allocation 2.42% Calibration 2.12% Wireless radio and link characteristics 2.12% Network monitoring 2.12% Geographic routing 1.82% Compression 1.82% Taxonomy 1.52% Capacity 1.52% Link-layer techniques 1.21% Topology control 1.21% Mobile nodes 1.21% Detection and estimation 1.21% Diffuse phenomena 0.91% Programming 0.91% Power control 0.61% Software 0.61% Autonomic routing 0.30%

2.5 SENSOR NODES

A sensor node, also known as a mote (produced in Berkeley, US), is a node in a WSN that is capable of performing some processing, gathering sensory information and communicating with other connected nodes in the network. One of the most constraints on sensor nodes is the low power consumption. Hence, sensor network protocols focus on power conservation. Since, the sensor nodes are often inaccessible, the lifetime of a sensor node must be assured. Lifetime of the sensor node depends on the lifetime of power resources. Power scarcity must be effectively managed.

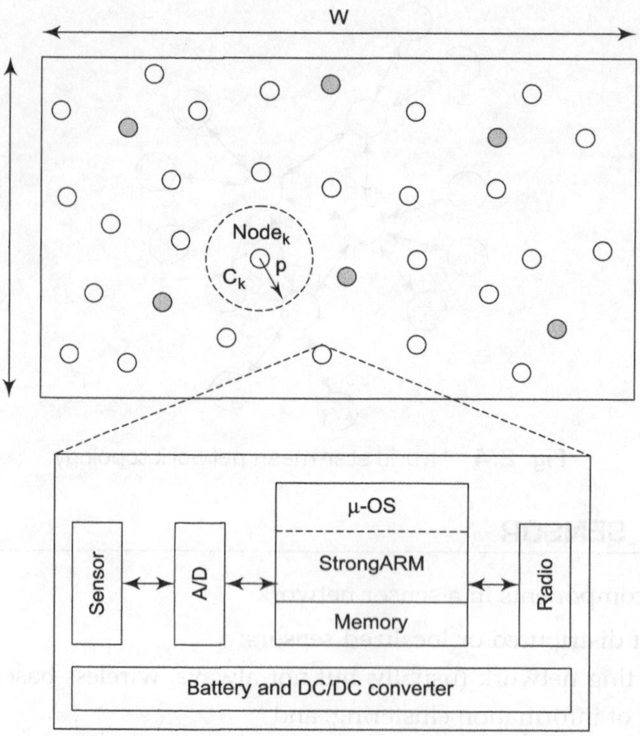

Fig. 2.5 The typical architecture of the sensor node

The architecture of the sensor node is shown in Fig. 2.5. It basically consists of - Sensing unit, computation or data processing unit and communication or radio unit. The wireless sensor nodes main components are a microcontroller, transceiver, external memory, power source and one or more sensors. Sensor data is converted to digital streams using ADC. Microcomputer unit (MCU) process this data streams by executing algorithms — such as classification algorithm. Then, the processed data is communicated over the network by radio unit.

Power consumption can be divided into three domains:

1. Sensing, 2. Communication, and 3. Data processing.

Sensing power varies with the nature of applications. Sporadic sensing might consume lesser power than constant event monitoring. Of the three domains, a sensor node expends maximum energy in data communication by radio unit. This involves both data transmission and reception.

I. Controller: The most common controller is a microcontroller. The controller performs tasks such as, processes data and controls the functionality of other components in the sensor node. A microcontroller is often used in many embedded systems sensor nodes because of its low cost, flexibility to connect to other devices, ease of programming, and low power consumption.

II. Transceiver: Sensor nodes possible choices of wireless transmission are Radio frequency (RF), Optical communication (Laser) and Infrared. WSNs tend to use license-free communication frequencies: 173, 433, 868, and 915 MHz; and 2.4 GHz. The transceiver functionality of both transmitter and receiver are combined in a single device. The operational states are transmitting, receiving, idle, and sleep modes.

The transceivers operating in idle mode consume power that is almost equal to the power consumed in receive mode.

III. External memory: Memory requirements are very much application dependent. Flash memories are used due to their cost and storage capacity. Two categories of memory used are: user memory and program memory.

IV. Power source: As most of sensors are wireless the power used are batteries or capacitors. The sensor node consumes power for sensing, communicating and data processing. The main source of power supply for sensor nodes batteries, both rechargeable and non-rechargeable. They are such as NiCd (nickel-cadmium), NiZn (nickel-zinc), Nimh (nickel-metal hydride), and lithiumion also the renewable energy from solar sources, temperature differences, or vibration.

V. Sensors: Sensors are hardware devices measure physical data of the parameter which is to be monitored. The wireless sensor nodes are typically very small electronic devices, and they are equipped with a limited power source of less than 0.5-2 ampere-hour and 1.2-3.7 volts.

Sensors are classified into three categories: passive, Omni-directional sensors; passive, narrow-beam sensors; and active sensors.

(*a*) Passive sensors sense the data without actually manipulating the environment by active probing. They are self powered; that is, energy is needed only to amplify their analog signals.

(*b*) Active sensors actively probe the environment, for example, a radar sensor, they require continuous energy from a power source.

(*c*) Narrow-beam sensors have a well-defined dimension of direction of measurement, similar to a camera.

(*d*) Omni-directional sensors have no notion of direction involved in their measurements. Omni-directional sensors and the spatial density of sensor nodes in the field may be as high as 20 nodes per cubic meter.

2.6 SENSOR ARCHITECTURE

Sensor nodes protocol layered network architectures is shown in Fig. 2.6. Each layer is responsible for the sensor to operate as per the role at each layer.

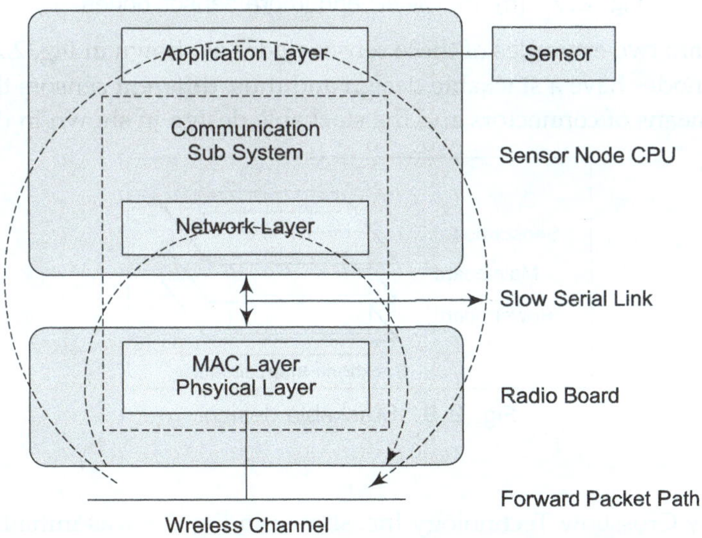

Fig. 2.6 Sensor architecture

I. The Applications Layer is responsible collecting the data from the application and forward the requests from the application layer down to the lower
II. The Transport Layer in conjunction with the Application layer does the computational control for the data to process in a controlled path. CPU controls the different states of the sensors for better energy efficient.
III. The Network Layer is responsible for routing information through the sensor network by finding the most efficient path for the packet to travel from source to sink..
IV. MAC protocol is responsible for medium access, error control, multiplexing of data streams and data frame detection. It ensures reliable point to point and point to multihop connections of the sensor nodes.
V. The Physical Layer is for the packet transfer in interference less mode. It is responsible for carrier frequency generation, signal detection, modulation and data packet transfer.
VI. Techniques used to reduce complexity and energy requirements, whilst improving reliability and reducing path loss effects and shadowing.

2.7 INDUSTRIAL SENSOR NODES

Industrial Sensor Nodes are the basic devices in wireless sensor networks. Such devices typically consist of a basic microcontroller, a radio, and a variety of sensors.

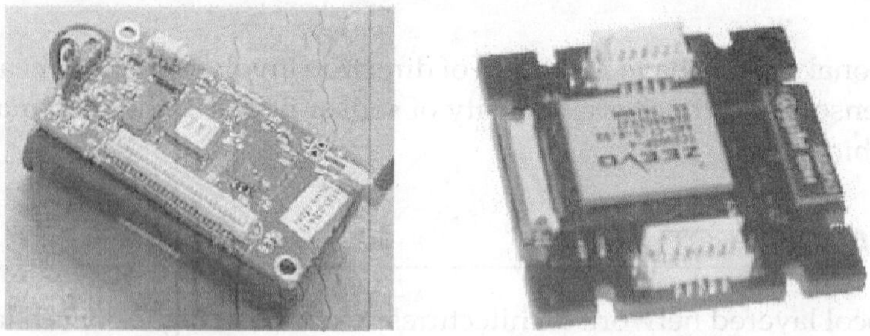

(a) Mica2 (b) Imote

Fig. 2.7 (a), (b) Mica2 and Imote sensor nodes.

Mica2 and Imote are two examples of these sensor nodes as shown in Fig. 2.7 (*a*) and (*b*) respectively. These sensor nodes have a stackable design and thus, different sensors that are piled up on the main board by means of connectors and the stackable design in shown in the Fig. 2.8.

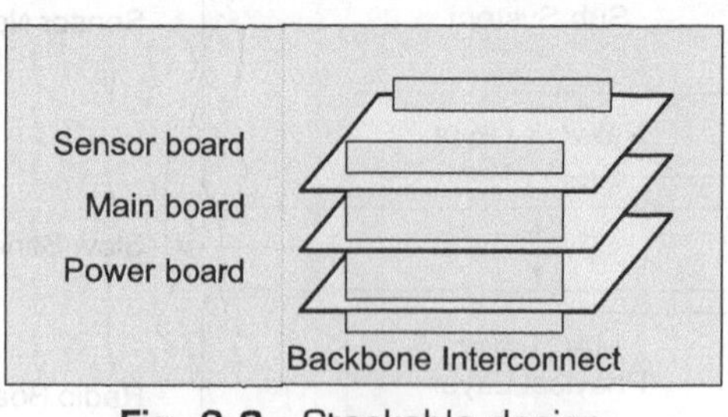

Fig. 2.8 Stackable design

2.7.1 Mica2

Mica2 is provided by Crossbow Technology Inc. shown in Fig. 2.9 was initially developed at the University of California, Berkeley.

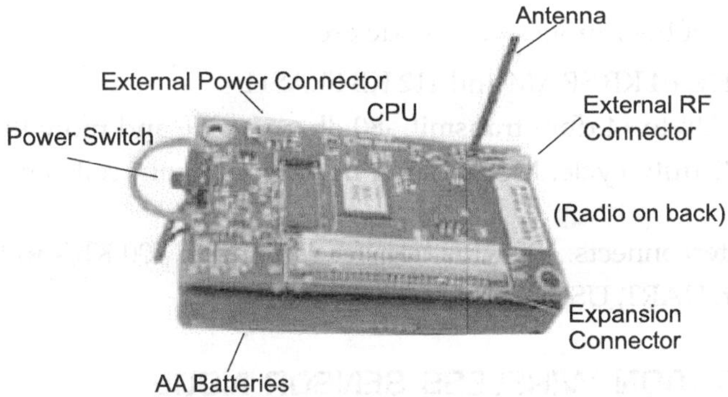

Fig. 2.9 Mica2 sensor node

The main components of a Mica2 sensor node include an antenna and a radio frequency (RF) transceiver to allow communication with other nodes, a memory unit, a CPU, the sensor unit (i.e., thermostat) and the power source usually by batteries. The operating system running on sensor nodes is called TinyOS. TinyOS is designed to run on platforms with limited computational power and memory space. The programming language of TinyOS is stylized C-Language and uses a custom compiler called NesC. It supported the platforms such as Linux RedHat 9.0, Windows 2000, and Windows XP.

The technical details of the Mica2 series (MPR4x0):

- It's a 3rd generation, tiny, wireless smart sensors.
- Operating system is TinyOS.
- Battery life is more than 1 year on AA batteries (using sleep modes).
- Router capability with wireless communications for every node.
- Multi-channel radio transceiver of 433, 868/916 or 310 MHz.
- RH, barometric pressure, temperature, acceleration/seismic, acoustic, magnetic, GPS and other sensors measurements.

2.7.2 I Mote

Imote technology is provided by Intel as shown in Fig. 2.10. I mote wireless communication is performed using the Bluetooth connectivity.

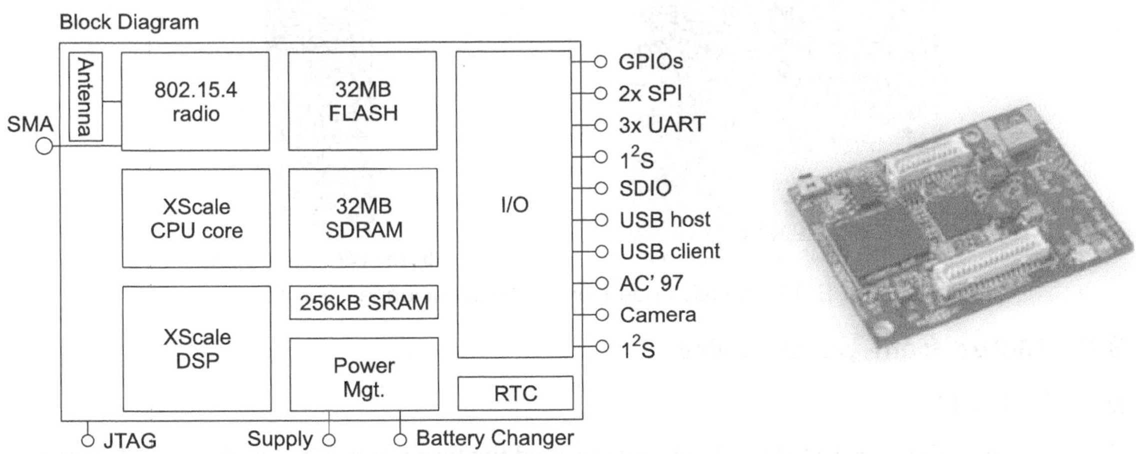

Fig. 2.10 I mote sensor node

The technical details of the I-mote sensor node are:

- ARM core: 12 MHz, 64 KB SRAM and 412 KB FLASH.
- Bluetooth radio: Up to +4 dbm transmit, -80 dbm receive and more than 30 m range.
- Battery life at 1% duty cycle: More than 1 month with coin cells and more than 6 months with AA cells.
- I2C backbone interconnects: 100 Kb/s transfer rate (up to 400 Kb/s in future revisions).
- Debug connector: UART, USB slave, JTAG.

2.8 NEXT GENERATION WIRELESS SENSOR NODE

2.8.1 Wireless Integrated Network Sensors (WINS)

In 1996, the Low Power Wireless Integrated Microsensors (LWIMs) were produced by UCLA at Rockwell Science Center. By using commercial, low cost CMOS fabrication, LWIMs demonstrated the ability to integrate multiple sensors, electronic interfaces, control, and communication on a single device. LWIM supported over 100 Kbps wireless communications at a range of 10 meters using a 1 mW transmitter.

In 1998, the same team built a second generation sensor node the 'Wireless Integrated Network Sensors' (WINS). Commercial WINS from Rockwell Science Center each consists of a processor board with an Intel Strong ARM SA1100 32-bit embedded processor of 1 MB SRAM and 4 MB flash memory, a radio board that supports 100 Kbps with adjustable power consumption from 1 to 100 mW, a power supply board, and a sensor board. The sensor boards are packaged in a 3.5″x3.5″x3″ enclosure as shown in Fig. 2.11 (*a*) and (*b*). The processor consumes 200 mW in the active state and 0.8 mW when sleeping mode of operation.

WINS offer relatively powerful processing and communication capabilities, other research efforts have been developing smaller and cheaper nodes with less power consumption. The figure shows the two boards of the WINS.

(a) The WINS processor board

(b) The WINS radio board

Fig. 2.11 WINS node from rockwell science center

2.8.2 Motes from UC Berkeley

2.8.2.1 WeC

In 1999, the Smart Dust project at UC Berkeley released the first node, WeC, in their product family of motes as illustrated in Fig. 2.12 (*a*). WeC was built with a small 8-bit, 4 MHz Atmel microcontroller

(512 bytes RAM and 8 KB flash memory), which consumed 15 mW active power and 45 μW sleeping power. WeC also had a simple radio supporting a data rate up to 10 Kbps, with 36 mW transmitting power and 9 mW receiving power. Later on, Rene and Dot were built in 1999 and 2000, respectively, with upgraded microcontrollers.

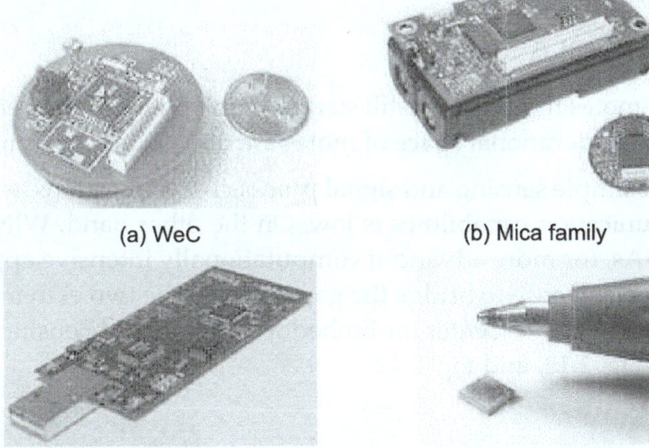

(a) WeC (b) Mica family

(c) Telos (d) Spec prototype

Fig. 2.12 Motes from UC Berkeley

2.8.2.2 Mica Family

The Mica family was released in 2001, it include Mica, Mica2, Mica2Dot, and MicaZ. Specifically, Mica was designed with 4 KB RAM, 128 KB flash, and a simple bit-level radio using RFM TR1000 that supported up to 40 Kbps with almost the same power consumption as the radio module on WeC.

Mote architecture allowed several different sensor boards to be stacked on top of the main processor/radio board. In the Fig. 2.12 (*b*) have shown the basic processor/radio board of approximate dimension of one inch by two inches in size.

The follow-ups to Mica, Mica2 and Mica2Dot were built in 2002 with an ATmega128L microcontroller that reduced standby current (33 mW active power and 75 μW sleep power).

2.8.2.3 Telos

The Fig. 2.12 (*c*) shows the latest member in the family. Telos, was released in 2004. It offered a set of new features:

I. A microcontroller from Texas Instruments with 3 mW active power and 15 μW sleep power,
II. An internal antenna built into the printed circuit board to reduce cost,
III. An on-board USB for easier interface with PCs,
IV. Integrated humidity, temperature, and light sensors, and
V. A 64-bit MAC address for unique node identification.

2.8.2.4 Spec

An interesting research tested is the Spec platform, which integrated the functionality of Mica onto a single 5 mm 2 chip is displayed in the Fig. 2.12 (*d*). Spec was built with a micro-radio, an analog-to-digital converter, and a temperature sensor on a single chip, which lead to a 30-fold reduction in total power consumption.

TinyOS features a component-based architecture and event driven model that are suitable for programming with small embedded devices, such as motes. The combination of Motes and TinyOS is gradually becoming a popular experimental platform for many research efforts in the field of WSNs.

2.8.3 Medusa from UCLA

With the development of motes the designer still start working on the energy efficient sensor nodes. The design philosophy and operational space of motes are quite different from those of WINS.

Motes are designed for simple sensing and signal processing applications, where the demand for computation and communication capabilities is low. On the other hand, WINS are essentially an embedded version of PDAs, for more advanced computationally intensive applications with large memory space requirements. Hence to bridge the gap between the two extremes, the Medusa MK-2 sensor node was developed by the Center for Embedded Networked Sensing (CENS) at UCLA in 2002 as illustrated in the Fig. 2.13, and Fig. 2.14.

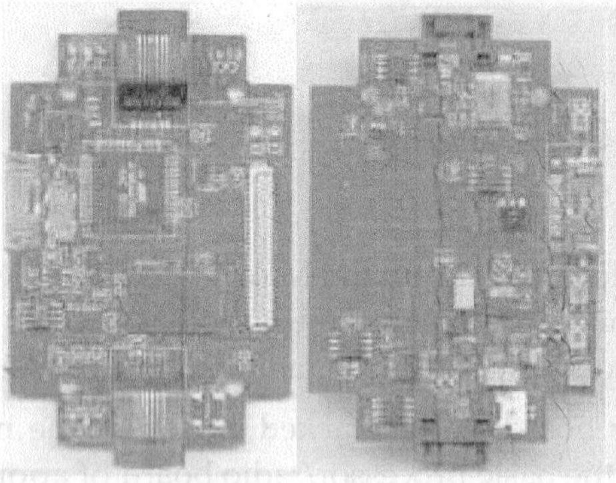

Fig. 2.13 Front and back side of board

Fig. 2.14 Medusa MK-2 sensor node from UCLA.

One distinguishing feature of Medusa MK-2 is that it integrates two microcontrollers.

- The first one, ATmega128 is dedicated to less computationally demanding tasks, including radio base band processing and sensor sampling.
- The second one, AT91FR4081, is a more powerful microcontroller (40 MHz, 1 MB flash, 136 KB RAM) that can be used to handle more sophisticated, but less frequent signal processing tasks (*e.g.*, the Kalman filter). The combination of these two microcontrollers provides more flexibility in WSN development and deployment, especially for applications that require both high computation capabilities and long lifetime.

2.8.4 Pico Radio from UC Berkeley

The above mentioned sensor architectures are based on batteries. Due to the slow advancement in battery capacity, techniques for energy scavenging from the environment there have been an attractive research field.

In 2003, the Berkeley Wireless Research Center (BWRC) presented the first radio transmitter, PicoBeacon as displayed in Fig. 2.15, is purely powered by solar and vibration energy sources.

Fig. 2.15 Pico Beacon from UC Berkeley

With a custom RF integrated circuitry that was developed for power consumption less than 400 μW, the beacon was able to achieve duty cycles up to 100% for high light conditions and 2.6% for typical ambient vibration conditions. It is anticipated that an integrated wireless transceiver with <100 μW power consumption is feasible in the near future.

The BWRC also produced SoC based sensor nodes instead of using COTS components. In 2002, PicoNode II was built using two ASIC chips that implemented the entire digital portion of the protocol stack. Together, the chip set consumed an average of 13 mW when three nodes were connected. The team is also building PicoNode III, which will integrate a complete PicoNode into a single small aspect-ratio package.

2.8.5 μAMPS from MIT

The μAMPS group from MIT the first test bed, μAMPS-I as shown in Fig. 2.16. The next development is μAMPS-II, where the developer is trying to build a highly integrated sensor node comprised of a digital and an analog/RF ASIC, μAMPS-II. The interesting feature of μAMPS-II is that the node will be able to operate in several modes. It can operate as a low-end stand-alone guarding node, a fully functional node for middle-end sensor networks, or a companion component in more powerful high-end sensor systems. Such nodes are energy efficient. Thus, it favours a network with heterogeneous sensor nodes for a more efficient utilization of resources.

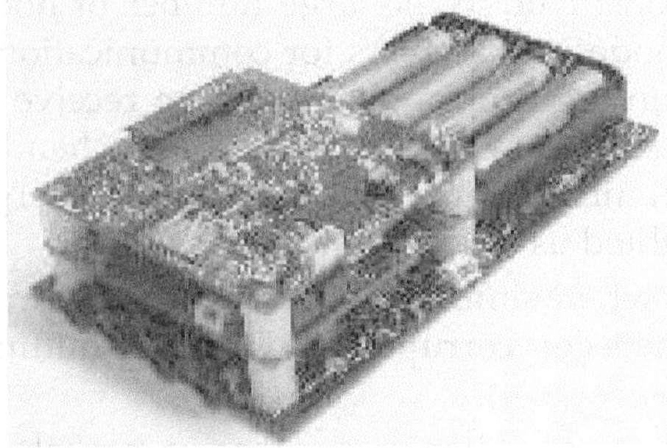

Fig. 2.16 Sensor node μAMPS from MIT

Besides the above sensor nodes, other commercial products and test beds for WSNs include Ember products, Sensoria WINS, Pluto mote, PC104 test bed, and Gnome test bed.

Table 2.1 illustrates the summary of motes. Imote2 is computationally powerful enough to run an embedded Linux kernel and requires a relatively lesser power supply (or a short usage period).

Table 2.1 Motes power summary

mote	processor	voltage	active	Sleep
Telos-B	IT MSP430	1.8V min	1.8 mA	5.1 uA
Mica-Z	Atmel AVR	2.5V min	8 mA	< 15 uA
Imote2	Intel PXA271	1.3V min	44-66 mA	390 uA

2.9 CHALLENGES OF SENSORS

Due to design issues sensors has a limited capabilities so, that it must be addressed to achieve an effective and efficient operation of wireless sensor networks. Some of the issues are highlighted as follows:

I. Energy constraints: Every sensor nodes have energy constraints. It is equipped with a limited power supply thus; its life time is heavily dependent on battery capacity. The sensing, transceiver and processing units all consume energy to carry out their tasks. However, most of the energy is consumed by the transceivers unit is for communicating data.

II. Processing limitation: The processing power of sensor nodes is directly linked to the amount of available energy in sensor networks. Although, the processing unit consumes less energy than the transmission unit of a node, that's why most wireless sensor nodes that run by low power micro-processors.

III. Energy saving algorithms: Since, sensor nodes use batteries to power the node, that are difficult to replace when drained, and it is not easy for replacement of the same at every case of applica-tion wherever installed. So, it is critical to design an algorithms and protocols to utilize minimal energy.

IV. Location discovery: Many applications can tracking an object require knowing the exact or approximate physical location of a sensor node in order to link sensed data with the object under investigation. Furthermore, many geographical routing protocols need the location of sensor nodes to forward data among the network. Location discovery protocols must be designed in such a way that minimum information is needed to be exchanged among nodes to discover their locations.

V. Communication limitations: Due to the large number of nodes and the potentially huge spatial spread between nodes transceivers for communication. The broadcast nature of radio leads all nodes within a given range are able to receive signal from the transmitting nodes. Due to the large numbers of nodes existence the bandwidth demand become large thereby causes the constraints. The four major sources of energy waste in such communication activities are highlighted as follow:

VI. Collision: This happens when simultaneously two nodes transmitting at the same time, thereby the received packets gets corrupted and the transmitting nodes need to be notifies to send the packet again.

VII. Overhearing: This mode occurs when nodes receive packets from other nodes that is not intended or addressed.

VIII. Overhead: The control packets serve various purposes that include reserving the medium for transmission, acknowledging and requesting of packets. This adds the energy cost of the nodes.

IX. Idle Listening: Apart from transmitting and receiving, a radio module consumes energy in the idle mode. In this mode, the transceiver is simply turned on but not transmitting or receiving.

X. Security: Secure routing, discovery and verification of location, key establishment and trust setup; attacks against sensor nodes, secure group management and secure data aggregation are some of the issues that address in a security context. Security solutions are constrained when applying them to sensor networks.

All sensor networks have above limitation, the constraints in energy is most important thereby reducing the life-span. Most sensor networks strive to minimize their energy usage in order to maximize their life time. This challenging task primarily involves in designing efficient communication and sensing protocols that significantly reduce energy leaks.

2.9.1 Sensor Characteristics

- Deeply distributed architecture: localized coordination to reach entire system goals, no infrastructure with no central control support.
- Autonomous operation: self-organization, self-configuration, adaptation.
- TCP/IP is open, widely implemented, supports multiple physical network, relatively efficient and light weight.
- Energy conservation: physical, MAC, link, route, application.
- Scalability: scale with node density, number and kinds of networks.
- Data centric network: address free route, named data, reinforcement-based adaptation, in-network data aggregation.

2.10 APPLICATIONS

Embedded built WSN covers a large range of application areas that overcome the challenges of wired network with the cost of power. As shown in Fig. 2.17 a WSN system is suitable for an application like remote monitoring, sustainability, industrial measurements etc. in the environmental monitoring it focus to acquire water, soil, or climate measurements, for utility like application focus towards electricity grid, street lights, and water location, health data. Wireless sensors offers lower cost and installed for effectively monitor highway, bridges and tunnels, airports, factories, power plant or production facilities etc.

The applications for WSNs are for specific applications, such as habitat monitoring, object tracking, fire detection, land slide detection and traffic monitoring also include for data collection.

Fig. 2.17 WSN application areas

I. Area monitoring: Area monitoring is a common application of WSNs. For example, collection of data from a large quantity of sensor nodes that is deployed in a battlefield to detect enemy intrusion, sensors for detection of event that is being monitored (heat, pressure, sound, light, electro-magnetic field, vibration, etc.), the event is reported to one of the base stations, to send a message via internet or to a satellite. Similarly, wireless sensor networks can use a range detection vehicles position.

II. Environmental Monitoring: A number of WSNs have been deployed for environmental monitoring. Because, some wireless sensor networks are easy to install, as per the needs of the application change.

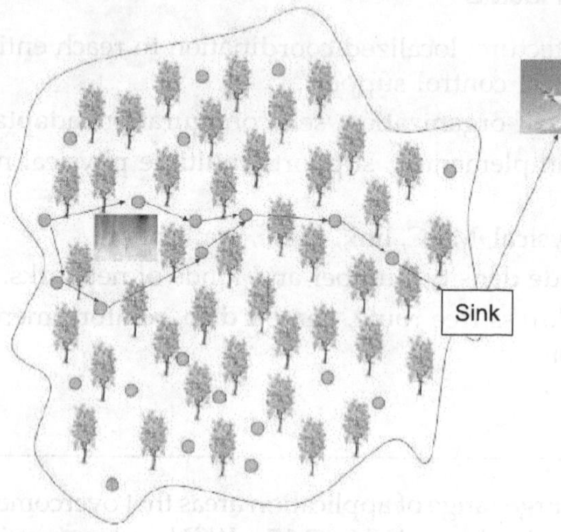

Fig. 2.18 Event detection application

As an example, environmental monitoring applications example, the location of a fire in a forest, or the detection of a quake, etc. can be monitors as per the Fig. 2.18.

III. Greenhouse Monitoring: Wireless sensor networks are also used to control the temperature and humidity levels. When the temperature and humidity parameter drops below specific levels, the greenhouse manager must be notified via e-mail or cell phone text message. The host systems can trigger misting systems, open vents, turn on fans, or control a wide variety of system responses.

IV. Landslide detection: A landslide detection system, make use of a wireless sensor network to detect the slight movements of soil that may occur during a landslide. And the data gathered it is possible to know the occurrence of landslides long before it actually happens.

V. Machine Health Monitoring: Wireless sensor networks have been developed for machinery condition-based maintenance as they offer significant cost savings and enable new functionalities. The application areas such as locations identification, rotating machinery, hazardous or restricted areas, etc can be reached with wireless sensors.

VI. Remote Medical Treatment: With the invention of internet technology the wireless sensor network jobs have been simplified. The sensors are placed in the various positions for the patient and through the remote location medical practitioners can able to diagnose the patient and monitor as displayed in Fig. 2.19.

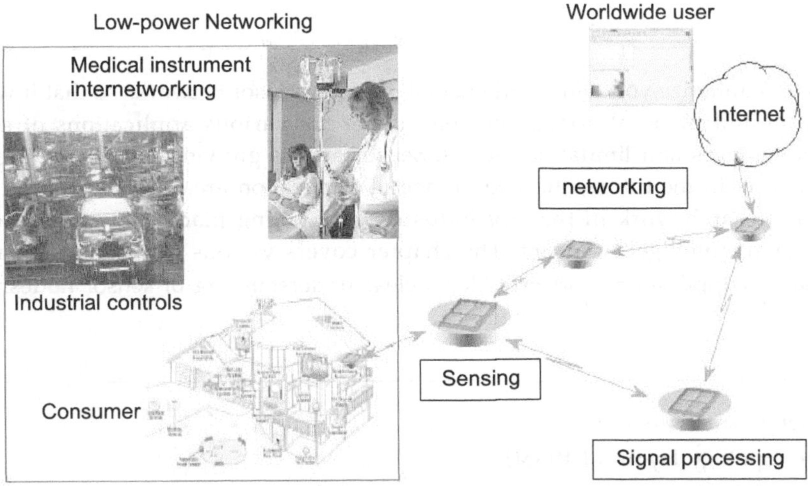

Fig. 2.19 Health monitoring/Industrial control

VII. Water/Wastewater Monitoring: There are many opportunities for using wireless sensor networks within the water/wastewater industries. The wireless data transmission can be monitored using industrial wireless I/O devices and sensors powered using solar panels or battery packs.

VIII. Landfill Ground Well Level Monitoring: Wireless sensor networks can be used to measure and monitor the water levels within all ground wells in the landfill site. The sensor information is transmitted wirelessly to a central data logging system to store the level data, perform calculations, as per application need.

IX. Water Tower Level Monitoring: Maintaining the water levels in these towers is important and requires constant monitoring and control. A submersible pressure sensors and float switches monitors the water levels in the tower and wirelessly transmits this data back to a control location. When the water levels fall, pumps discharge more water from the reservoir to the tower are turned on.

X. Agriculture: Using wireless sensor networks within the agricultural industry is increasingly common. Base on water requirements data is transmitters to monitor water tank levels, pumps can be controlled using wireless I/O devices, and water that use can be measured and wirelessly transmitted back to a central control center for billing. Irrigation automation enables more efficient water use and reduces waste.

XI. Fleet monitoring: It is possible to put a mote with a GPS module on-board of each vehicle of a fleet. The mote gathers its position via the GPS module, and reports its coordinates so that the location is tracked in real-time. The mote is equipped with temperature sensors to avoid any disruption of the cold chain, helping to ensure the safety of food, pharmaceutical and chemical shipments.

XII. Future Developments: The most general and versatile deployments of wireless sensing networks demand that batteries be deployed. Future work is being performed on systems that exploit piezoelectric materials to harvest ambient strain energy for energy storage in capacitors and/or rechargeable batteries. By combining smart, energy saving electronics with advanced thin film battery chemistries that permit infinite recharge cycles, these systems could provide a long term, maintenance free, wireless monitoring solution.

SUMMARY

This chapter was to familiarize you with fundamental issues of sensor networks so that it will give ideas on the sensor, its form and physical appearance etc. There are various applications of sensor networks followed by its advantages and limitation factors. Sensor area is growing very fast, attracting more and more people to its use. In the future, the wide range of application area will depend on wireless sensor networks. Several research work in the lab/ industries are taking place to design an energy efficient sensors towards smart intelligent sensors. The chapter covers various form of sensor nodes currently employs in its various applications and provides a clear understanding of sensor nodes.

QUESTIONS

1. Explain briefly, what is WSN?
2. What is the topology layout of WSN?
3. Draw and explain the internal structure of sensor nodes.
4. What are the different types of Industrial sensor nodes?
5. State the different challenges of sensors.
6. Illustrates the application of sensor nodes.

BIBLIOGRAPHY

- Aguayo D, Bicket J, Biswas S, Judd G, Morris R (2004) Link level measurements from an 802.11b mesh network. Proceedings of ACM Sigcomm, pp 121–131, USA, August
- Perkins CE, Royer EM (1999) Ad-hoc on-demand distance vector routing. In: Proc. of IEEE workshop on mobile computing systems and applications, New Orleans, LA, February 1999.
- Gong MX, Midkiff SF, Mao S (2005) Design principles for distributed channel assignment in wireless Ad-Hoc networks. In: Proc. of IEEE ICC, Seoul, and May 2005.
- IEEE (1999) IEEE standard for wireless LAN-medium access control and physical layer specification. IEEE 802.11 Std.
- Krishnamurthy S, Thoppian M, Venkatesan S, Prakash R (2005) Control channel based MAC-Layer configuration, routing and situation awareness for cognitive radio networks. In: Proc. of IEEE MILCOM, Atlantic, NJ, and October 2005.
- Bianchi G (2000) Performance analysis of the IEEE 802.11 distributed coordination function. IEEE J Sel Areas Commun 18:535–547
- Van den Berg J, Roijers F, Mandjes M (2006) Performance modeling of a bottleneck node in an IEEE 802.11 Ad-hoc network. In: The 5th International Conference on Ad-hoc Networks and Wireless, Ottawa, August 2006.
- IEEE Std 802.11a-1999 (1999) Wireless LAN medium access control (MAC) and physical layer (PHY) specifications: high speed physical layer in the 5 GHZ band
- Jain K, Padhye J, Padmanabhan VN, Qiu L (2003) Impact of interference on multi-hop wireless network performance. Proc. of ACM Mobicom, p. 66–80
- Tebbe H, Kassler A, Ruiz P (2006) QoS-aware mesh construction to enhance multicast routing in mobile ad-hoc networks. Proceedings of intersense 2006, May.
- Rate-Diversity and Resource-Aware Broadcast and Multicast in Multi-rate Wireless Mesh Networks, Bao Hua Liu, Chun Tung Chou, Archan Misra and Sanjay Jha, Mobile Netw Appl (2008).

3

Layers in Wireless Sensor Network

3.1 INTRODUCTION

With the development of large scale integrated circuits and wireless communication technologies, the application prospective of wireless sensor networks (WSNs) by interconnected tiny sensor nodes becomes more and more attractive in various areas. Wireless sensor networks constitute a special class of wireless data communication networks. The networks of wireless sensor nodes are distributed in a region. Each node is dynamic has a limited energy supply generates information to be communicated to a sink node. The transmission power, modulation scheme, and duty cycle of node varies in real time is dynamic in nature. So, the computation of optimal transmission powers, rates, and link schedule that maximize the network lifetime observe periodically.

Communication in wireless sensor networks is divided into several layers. The WSN protocols for accessing the shared communications medium have design objectives which are quite different from those used in other types of computer networks. However, due to the advancement in computing towards ubiquitous computing applications the technologies are yet to update to solve some of the issues such as:

(*i*) *Network lifetime:* The long lifetime requirement of applications depends upon the battery capacity and the node power consumption. So, based on power-aware protocols, low power hardware design, power-saving sleep mode, and transmission range optimization network life can be increase.

(*ii*) *Scalability:* WSN application contains thousands, if not millions, of sensor nodes. The cost of such systems is inversely proportional to size of applications and directly proportion to unit price per sensor node if it is low enough. The affordability requirement imposes budget constraints on hardware components at each sensor node that directly affect its communication bandwidth, computing/storing capabilities. This also brings challenges to protocols' design for WSN so that their implementations can be scalable and simple.

(*iii*) *Functionality:* In WSN nodes take multiple responsibilities of collecting various types of data, processing and fusing data and thereby relaying data via multi-hop transmission to improve communication efficiency. This multi-responsibility requirement is challenging to the resource-constrained sensor node, and issues of scalability and network lifetime.

3.2 ISSUES IN WIRELESS SENSOR NETWORKS

With the development of large scale integrated circuits and wireless communication technologies, the application prospective of wireless sensor networks (WSN) by interconnected tiny sensor nodes becomes more and more attractive.

(i) *Scalability*: WSN contain thousands, if not millions, of sensor nodes depending on applications. The cost of such systems will be unaffordable if the unit price per sensor node is not low enough. The affordability requirement imposes budget constraints on hardware components at each sensor node that directly affect its communication bandwidth, computing/storing capabilities, and availability of specific devices such as GPS receivers. This also brings challenges to protocols' design for WSN so that their implementations can be scalable and simple.

(ii) *Network lifetime*: The long lifetime requirement of applications and the limited capacity of batteries create a wide gap between the node power consumption and the node power supply. Current solutions include power-aware protocols, low power hardware design, power-saving sleep mode, and transmission range optimization.

(iii) *Functionality*: In WSN some nodes are expected to take multiple responsibilities of collecting various types of data, processing and fusing data to improve communication efficiency, and relaying data via multi-hop transmission. This multi-responsibility requirement is challenging to the resource-constrained sensor node that are above-mentioned issues of scalability and network lifetime.

3.3 WIRELESS SENSOR NETWORK PROPERTIES

Wireless Sensor Network is a network of wireless embedded system elements, which consists of spatially distributed autonomous devices using sensors to cooperatively monitor physical or environmental conditions, such as temperature, sound, vibration, pressure, motion or pollutants at different locations. WSNs belong to the Low-Rate Wireless Personal Area Network (LR-WPAN). Today wireless communication becomes more and more popular and a large variety tending from Wireless Personal Area Networks (WPAN), with ranges from 10 meters to Wireless Regional Area Networks (WRAN), extending even up to 40 kilometers, and Satellite Communication, allow to build any specialized or general purpose network.

3.3.1 The Following Properties, which Describe WSNs, are:

(i) ad-hoc/coordinated communication, distributed organization,

(ii) mobility of nodes, dynamic network topology, source-sink structure, variable density,

(iii) Intermittent connectivity, node failures, large area of network deployment.

3.3.2 A Wireless Sensor Node is an Element of a WSN, Having Such Properties

(i) small size, battery powered,

(ii) low power consumption, low data rates, low cost,

(iii) Sensing/monitoring equipment, intelligent structure.

A sensor network is defined as being composed of a large number of nodes which are deployed densely in close proximity to the phenomenon to be monitored. Each of these nodes collects data and route this information back to a sink. The network must possess self-organizing capabilities since the positions of individual nodes are not predetermined.

Major differences between sensor and ad-hoc networks are:

(*i*) Number of nodes can be orders of magnitude higher.
(*ii*) Sensor nodes are densely deployed.
(*iii*) Sensor nodes are prone to failure.
(*iv*) Frequent topology changes.
(*v*) Broadcast communication paradigm.
(*vi*) Limited power, processing and power capabilities.
(*vii*) Possible absence of unique global identification per node.

3.4 LAYERS OF OSI MODEL

The International Standards Organization (ISO) Open Systems Interconnect (OSI) is a standard set of rules describing the transfer of data between each layer. Each layer has a specific function. The OSI Model clearly defines the interfaces between each layer. This allows different network operating systems and protocols to work together to the standard interfaces. The application of the ISO OSI model has allowed the modern multi protocol networks that exist today. The seven layers of the OSI reference model can be divided into two categories: upper layers and lower layers as shown in Fig. 3.1.

There are 7 Layers of the OSI model:

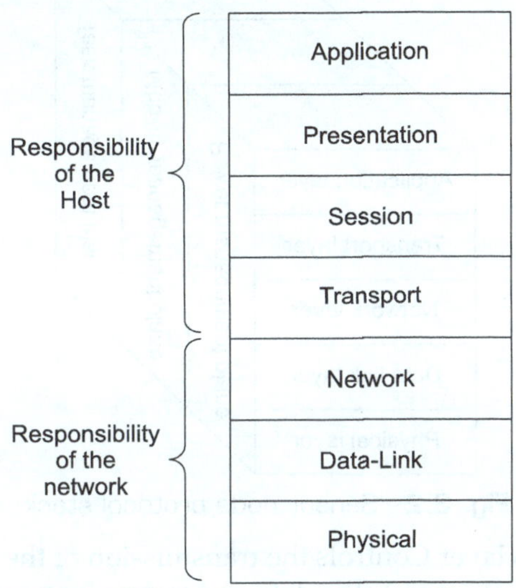

Fig. 3.1 The layers make up the OSI layers

- Application Layer (Top Layer)
- Presentation Layer
- Session Layer
- Transport Layer
- Network Layer
- Data Link Layer
- Physical Layer (Bottom Layer)

Basically in the OSI Model, we have two layers:

(*i*) Upper Layer – Comprising of Sessions, Presentation and Applications Layers.
This layer basically deals with the application issues (basically software). Application Layer is the closest to the end user. For eg: Applications like Microsoft Word, PowerPoint to be closer to the end user like us.

(*ii*) Lower Layer – Comprising of the Physical, Data Link, Network and Transport Layers.
This layer deals with the data transport issues. The Physical and Data Link layers basically deal with hardware and software. Other Lower Layers are generally implemented using software.

3.5 PROTOCOLS

The OSI model provides a conceptual framework for communication between two computer devices. The communication is made possible by using communication protocols. The protocol is a formal set of rules and conventions that governs how computers exchange information over a network medium. A protocol implements the functions of one or more of the sensor nodes adhere to a specific protocol stack as shown in Fig. 3.2. Layered network architecture adopted because they most certainly always improve the robustness of a system. A lot of research is still being conducted in perfecting the protocol stack for sensor network, so the exact protocols are yet to be concreted.

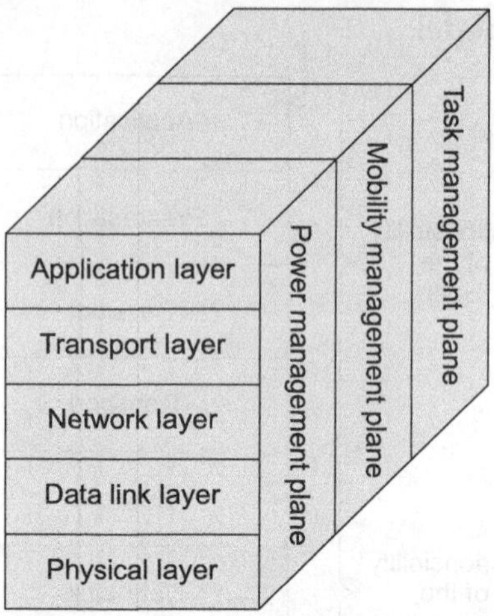

Fig. 3.2 Sensor node protocol stack

1. The Physical Layer: This layer Controls the transmission of the actual data onto the network nodes. The electrical signals encode the data and transmit over network. Layer responsible is for carrier frequency generation, frequency selection, signal detection, modulation and data encryption. Various techniques such as Ultra Wideband, Impulse Radio and Pulse Position modulation have been used to reduce complexity and energy requirements, whilst improving reliability and reducing path loss effects and shadowing.
2. The Data Link Layer: This layer takes the data frames or messages from the Network Layer and provides for their actual transmissions. Layer is responsible for medium access, error control, multiplexing of data streams and data frame detection. It is to ensure reliable point to point and point to multihop connections in the network. Due to the network constraints, conventional MAC protocols are not suited to sensor networks.

3. The Network Layer: This layer is responsible for addressing messages and data so they are sent to the correct destination, and for translating logical addresses and names into physical addresses. This layer is also responsible for finding a path through the network to the destination computer. Layer is responsible for routing information through the sensor network for finding the most efficient path for the packet to travel from source to sink.
4. The Transport Layer: This layer ensures that data is delivered error free, in sequence and with no loss, duplications or corruption. This layer repackages data into smaller messages for sending, and repackaging into the original larger message at the receiving end. Layer is needed when the sensor network intends to be accessed through the internet. Since, nodes are dynamic fixed IP are not compatible. Modified TCP/UDP like protocols may be an appropriate solution where several study is going on.
5. The Applications Layer: This layer provides Applications with access to network services. Its responsible presenting all required information to the application and propagating requests from the application layer down to the lower layers.

3.6 ISSUES OF DATA LINK LAYERS

The goal of the data link layer is to provide reliable, efficient communication between adjacent machines connected by a single communication channel.

3.6.1 Design Issues

Below are the some of the important design issues of the data link layer:

(*i*) Reliable Delivery: Frames are to be delivered to the receiver reliably and in the same order as generated by the sender. Connection state need to keep track of sending order and which frames that require retransmission.

(*ii*) Best Effort: The receiver does not return acknowledgments to the sender, so the sender has no way of knowing if a frame has been successfully delivered. How the higher layers can recover from errors and ease to have higher layers deal with occasional loss of packet.

(*iii*) Acknowledged Delivery: The receiver returns an acknowledgment frame to the sender indicating that a data frame was properly received but may not necessarily retransmit unacknowledged frames. So, the design should distinguish the missing packet acknowledgement over the transmission of next packets.

(*iv*) Framing: It is ensure the receiver detect frame boundaries such as Length Count, Bit Stuffing, and Character stuffing.

(*v*) Error Control: The protocol should concern on error control by insuring that all frames are eventually delivered to a destination. Hence, there need for proper updating on Acknowledgements, Timers, and Sequence Numbers.

(*vi*) Flow Control: Flow control deals with throttling the speed of the sender to match that of the receiver in a dynamic process, as the receiving speed depends on such changing factors as the load, and availability of buffer space.

3.7 ISSUES OF NETWORK LAYERS

The network layer is to provide an end-to-end communication capability in contrast to machine-to-machine communication provided by the data link layer. This end-to-end is performed using two basic approaches known as connection- oriented or connectionless network-layer services. Below are the some of the important design issues of the network layer:

(*i*) Interface between the host and the network
(*ii*) Routing
(*iii*) Congestion and deadlock
(*iv*) Internetworking

3.7.1 Network Layer Interface

There are two basic approaches used for sending packets, which is a group of bits that includes data plus source and destination addresses, from node to node called virtual circuit and datagram methods. They are also referred to as connection-oriented and connectionless network-layer services.

3.7.2 Overview of Other Network Layer Issues

The network layer is responsible for routing packets from the source to destination. The routing algorithm is the piece of software that decides, where a packet goes next (*e.g.*, which output line, or which node on a broadcast channel).

For connectionless networks, the routing decision is made for each datagram. For connection-oriented networks, the decision is made once, at circuit setup time.

3.7.3 Routing Issues

The routing algorithm must deal with the following issues:

(*i*) Correctness and simplicity: Networks are never taken down; individual parts (*e.g.*, links, routers) may fail, but the whole network should not.
(*ii*) Stability: If a link or router fail time elapses before the remaining routers recognize the topology change.
(*iii*) Fairness and optimality: The channel and delay shall be minimum and sustain the network.

3.7.4 Congestion

The network layer also must deal with congestion:

(*i*) When packets quantity is more than processing capacity, delays increase and performance decreases. If the situation continues, the subnet discard packets.
(*ii*) If the delay increases, the sender retransmit, making a bad situation even worse.
(*iii*) Overall, performance degrades because the network loose resources processing packets that eventually get discarded.

3.7.5 Internetworking

Internetworking is the connection of different network technologies together. The problems are:

(*i*) Packets may travel through many different networks
(*ii*) Each network may have a different frame format
(*iii*) Some networks may be connectionless, other connection oriented

3.8 ISSUES OF TRANSPORT LAYERS

The transport level provides end-to-end communication between processes executing on different machines are similar to those provided by a data link layer protocol, there are several important differences between the transport and lower layers:

(*i*) Quality and Type of Services: The user and transport protocol need to negotiate as to the quality or type of service to be provided optimum such as throughput, delay, protection, priority, reliability, etc.

(*ii*) Guarantee Service: The transport layer may have to overcome service deficiencies of the lower layers (e.g. providing reliable service over an unreliable network layer).

(*iii*) Deal with congestion control: In connectionless Internets, transport protocols must exercise congestion control. During network congestion, they must reduce rate at which they insert packets into the subnet, because the subnet has no way to prevent itself from becoming overloaded.

(*iv*) Connection establishment: Transport level protocols go through three phases: establishing, running, and terminating a connection. For data gram-oriented protocols, opening a connection simply allocates and initializes data structures in the operating system kernel.

3.9 NETWORK CHARACTERISTICS

A WSN typically consists of a large number of low-cost, low-power, and multi-functional sensor nodes that are deployed in a region of interest. These sensor nodes are small in size, but are equipped with sensors, embedded microprocessors, and radio transceivers, and therefore have not only sensing capability, but also data processing and communicating capabilities. Compared with traditional wireless communication networks, for example, cellular systems and MANET, sensor networks have the following unique characteristics and constraints:

(*i*) Dense Node Deployment: Sensor nodes are usually densely deployed in a field of interest. The number of sensor nodes in a sensor network can be several orders of high magnitude.

(*ii*) Battery-Powered Sensor Nodes: Sensor nodes are usually powered by battery. In most situations, they are deployed in a harsh or hostile environment, where it is very difficult or even impossible to change or recharge the batteries.

(*iii*) Severe Energy, Computation, and Storage Constraints: Sensor nodes are highly limited in energy, computation, and storage capacities.

(*iv*) Self-Configurable: Sensor nodes are usually randomly deployed without careful planning and engineering. Once deployed, sensor nodes have to autonomously configure themselves into a communication network.

(*v*) Application Specific: Sensor networks are application specific. A network is usually designed and deployed for a specific application. The design requirements of a network change with its application.

(*vi*) Unreliable Sensor Nodes: Sensor nodes are usually deployed in harsh or hostile environments and operate without attendance. They are prone to physical damages or failures.

(*vii*) Frequent Topology Change: Network topology changes frequently due to node failure, damage, addition, energy depletion, or channel fading.

(*viii*) No Global Identification: Due to the large number of sensor nodes, it is usually not possible to build a global addressing scheme for a sensor network, because it would introduce a high overhead for the identification maintenance.

(*ix*) Many-to-One Traffic Pattern. In most sensor network applications, the data sensed by sensor nodes flow from multiple source sensor nodes to a particular sink, exhibiting a many-to-one traffic pattern.

(*x*) Data Redundancy. In most sensor network applications, sensor nodes are densely deployed in a region of interest and collaborate to accomplish a common sensing task. Thus, the data sensed by multiple sensor nodes typically have a certain level of correlation or redundancy.

These unique characteristics and constraints present many new challenges in the design of sensor networks.

3.10 NETWORK AND PROTOCOL ISSUE

In the networking and communication it is important that the sensor network nodes have to be networked and communicated with each other in order to do its objective job in a collaborative manner. Many research practices are currently engaged in developing the technologies needed for different layers of the sensor networks protocol stack.

3.10.1 The Physical Layer

The Open research issues in this layer are:

(*i*) Power Efficiency Modulation Schemes: M-array, Binary Modulation scheme, Ultra wide, impulse radio and position modulation (PPM) are used for the power efficient purpose

(*ii*) Strategies to overcome signal propagation effects: The minimum output power required to transmit a signal over a distance d is proportional to dn, where 2<= n < 4. So to reduce the output power requirement of sensor node various strategies have to be explored. Usually, multihop communication is followed for this purpose in a sensor network; it can effectively overcome shadowing and path loss effects

3.10.2 The Data Link Layer

Open research issues in DLL:

(*i*) MAC for mobile sensor networks: MAC establishes links; fairly and efficiently share communication resources. It must have built-in power conservation, mobility management, and failure recovery strategies.

(*ii*) The adaptive transmission rate control (ARC): Achieves medium access fairness by balancing the rates of originating and route-through traffic. It controls the data origination rate of a node in order to allow the route-through traffic to propagate. The route-through traffic is preferred over the originating traffic. It follows linear increase and multiplicative decrease approach, since dropping route-through traffic is costlier, the associated penalty is lesser

(*iii*) Error control coding schemes: Simple error control codes with low-complexity encoding and decoding might present the best solutions for sensor networks.

(*iv*) Power-saving modes of operation: The issues in this topic involved are focusing on how to save power by turning off the transceiver when it is not required. The Operation in a power saving mode is energy efficient only if the time spent in that mode is greater than a certain threshold.

3.10.3 Network Layer

The layer responsibilities are routing with Power efficiency, addressing and location awareness.

(*i*) Flooding: Each node receiving a packet repeats it by broadcasting. It does not require costly topology maintenance and complex route discovery algorithms.

(*ii*) Implosion: a situation where duplicated messages are sent to the same node.

(*iii*) Overlap: If two nodes share the same observing region, both of them may sense the same stimuli at the same time. As a result, neighbor nodes receive duplicated messages.

(*iv*) Resource blindness: Flooding does not take into account the available energy resources.

(*v*) Gossiping: No broadcast, but send the incoming packets to a randomly selected neighbor. Avoids the implosion problem. It takes a long time to propagate the message to all sensor nodes.

(*vi*) Data aggregation problem: A technique used to solve the implosion and overlap problems in data-centric routing. Data coming from multiple sensor nodes with the same attribute of phenomenon are aggregated. With this respect, data aggregation is known as data fusion.

3.10.4 Transport Layer

This layer is function is to be accessed through the Internet or other external networks. Usually, the between sink node and user UDP or TCP is used.

3.10.4.1 Open Research Issues

The development of transport layer protocols is a challenging effort with the Hardware constraints such as the limited power and memory. There is no possibility to have buffer like TCP, acknowledgements are too expensive.

The protocols in these layers are Sensor management protocol, Task assignment and data advertisement protocol, Sensor query and data dissemination protocol, Traffic Management and Admission Control protocol to prevent applications from establishing data flows when the network resources needed are not available.

The potential application areas for sensor networks defines and propose, potential application layer protocols for sensor networks remains a largely unexplored region need to be developed to provide a greater level of services. So, there several research studies carry out on pure application layer protocols.

3.11 NETWORK TOPOLOGIES

Depending on the application requirements, the WPAN may operate in one of three topologies:

(*i*) The star topology,

(*ii*) The peer-to-peer topology

(*iii*) Cluster tree topology.

3.11.1 The Star Topology

In star topology, as shown in Fig. 3.3, all nodes are connected to a central point which is called a hub. A sending node transmits its message (signal) to the hub. The hub then retransmits the signals to all other nodes. Each receiving node checks the signal's destination address; if it matches the node's address it processes the signals.

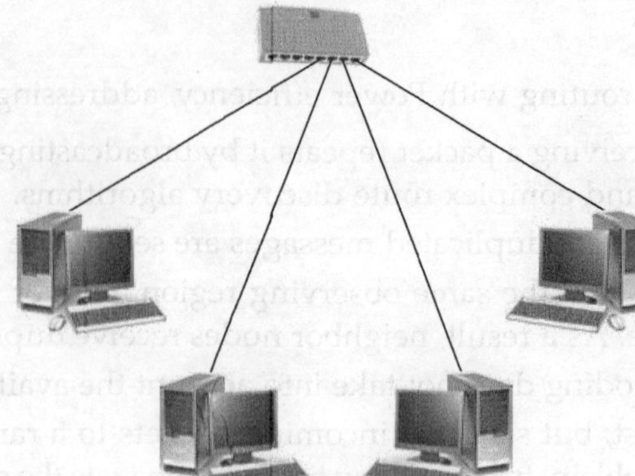

Fig. 3.3 Star topology

Figure 3.4 shows that the star topology, where the communication is established between end devices and a single central controller, which is called the Personal Area Network (PAN) coordinator. A PAN coordinator can be used to start, stop or route communication in the network. The PAN coordinator is the primary controller of the PAN, which can be mains or battery powered.

Applications that benefit from a star topology are home automation, personal computer peripherals, toys and games and personal health care.

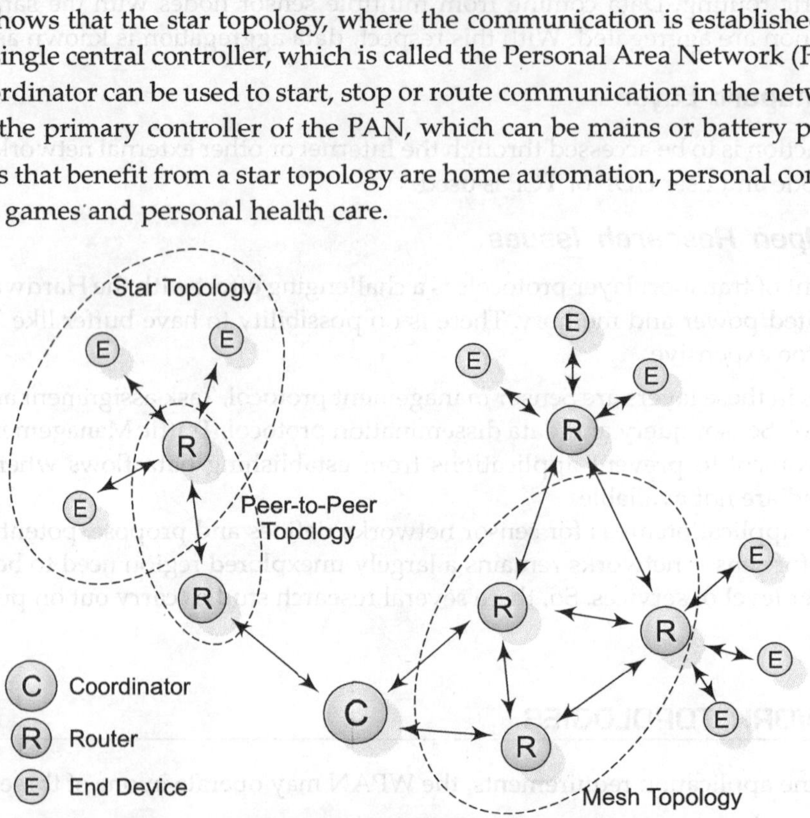

Fig. 3.4 Star and peer-to-peer network topologies

3.11.2 The Peer-to-Peer Topology

Peer-to-peer (P2P) forms a network of nodes when they are connected directly. The Fig. 3.5 shows the mesh network connectivity of nodes with several links directly. The peer-to-peer topology also includes the PAN Coordinator or router. However, it is different from the star topology.

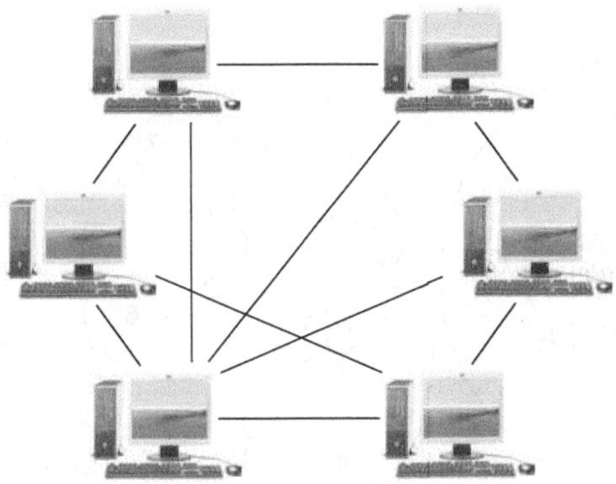

Fig. 3.5 Peer-to-Peer network

The difference is that devices can communicate directly as long as they are in communication range of each other. The peer-to-peer topology allows more complex network formation, such as mesh networking. A peer-to-peer network can be ad-hoc and self-organizing. The network structure allows multiple hops from any device to any other device on the network for message routing.

Application of peer-to-peer topology is done in the areas of industrial control and monitoring, logistics, tracking, intelligent agriculture and security applications.

3.11.3 Cluster Tree Topology

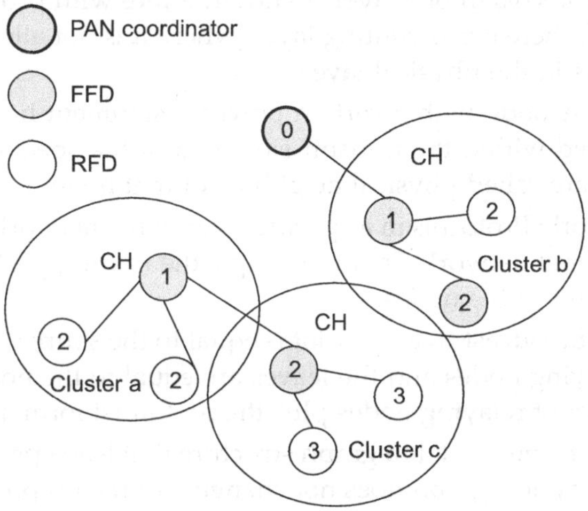

Fig. 3.6 Cluster tree topology

In large WSNs, for scalability reasons, constituted by hundreds of sensors nodes may refer to multiple sinks or may organize themselves in clusters tree topology as shown in Fig. 3.6. The Cluster Heads (CH), role is to collect the data of the cluster and route them towards the sink via a hop-by-hop path on other CHs. The topology is called cluster-tree.

The cluster-tree network topology as shown in Fig. 3.7 is a special case of a peer-to-peer network. Clusters are interconnected by a leaf node that appears at the cluster edge. The cluster-tree networks are constructed.

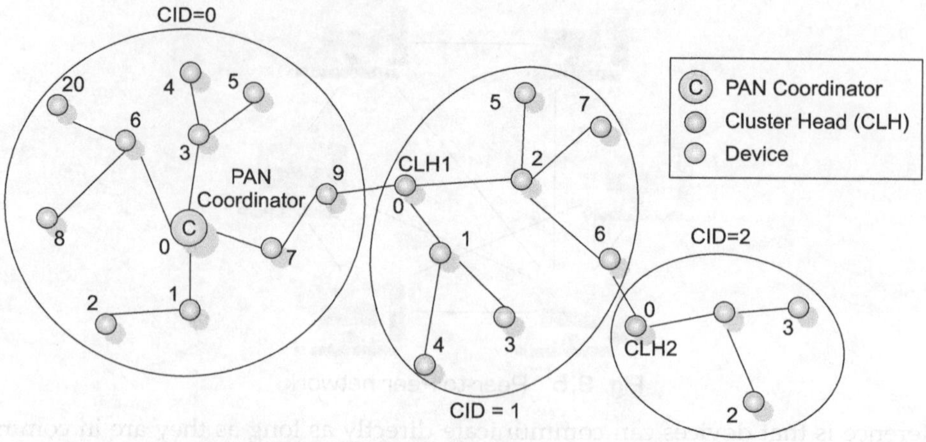

Fig. 3.7 Cluster-tree network topology

3.12 NETWORK AREA CLASSIFICATION

Network area classification can be analyzed to increase the network life time. To categorize the Network Lifetime Problems there are several terms from network of network classification. Some of the network classification is a follows:

I. Layered Network: In a layered network, every layer are distinguished. The network layers are separated, *i.e.* protocols in one layer do not interfere with protocols in another layer. In the routing protocol, there exists routing layer, where it is not allowed to control the power assignment, which is in the physical layer.

II. Physical neighbors: A node with a certain power assignment has a transmission area. All nodes that are located within the transmission area of the specific node can communicate with that node and are called physical neighbors of that node.

III. Cross-layered Network: Protocols in a specific layer of the network may interfere with other layers in a cross-layered network. So, for example the routing protocol is allowed to change the power assignments of the nodes.

IV. Broadcast Tree: In a broadcast tree, the root is equal to the source node, the non-leave nodes are equal to the relaying nodes and the leaves are equal to the non-relaying nodes. To be a broadcast tree, the set of relaying nodes plus the root must form a Connected.

V. Logical Neighbors: The network is a graph structure that leads power assignments to physical neighbors. The physical graph does not change, but the graph that is used for computational is called the logical graph. The neighbors of a node in the logical graph are called logical neighbors. Obviously, all logical neighbors of a node are also physical neighbors of the node. Furthermore, a node can have physical neighbors than logical neighbors.

3.13 NETWORK DESIGN CHALLENGES

WSN important issues related to node-level are limited resource management, concurrency handling, power management and memory management, where as issues related to both are internode communication, failure handling, heterogeneity and scalability. A sensor node is constrained by the resources available to it is constrained by limited battery power, processing capability, memory and bandwidth.

The unique network characteristics present many challenges in the design of sensor networks, which involve the following main aspects:

I. Limited Energy Capacity: Sensor nodes are battery powered and thus have very limited energy capacity. This constraint presents many new challenges in the development of hardware and software, and the design of network architectures and protocols for sensor networks. To prolong the operational lifetime of a sensor network, energy efficiency should be considered in every aspect of sensor network design, not only hardware and software, but also network architectures and protocols.

II. Battery Power: Power consumption is crucial to the life span of WSN based applications. Most of the applications in WSN a typical node with a limited power supply have to live mostly for months to years. The main source of power consumption is communication when compared to computation and sensing.

   (*i*) The energy cost of transmitting a bit of data over RF channel is equivalent to executing thousands of instructions by the processor on the node. Reading and writing to flash storage also consumes significant amount of energy consumption overhead while loading and unloading of modules into program memory.

   (*ii*) Sensor nodes operate in three sleep modes, idle, power down and power save, in order to conserve the energy. In idle mode the processor alone shuts off, power down mode shuts off everything except the watch dog timer and interrupt logic necessary to wake up the node, power save mode is similar to power down mode except that it keeps the timer running.

III. Limited Hardware Resources: Sensor nodes have limited processing and storage capacities, and thus can only perform limited computational functionalities. These hardware constraints present many challenges in software development and network protocol design for sensor networks, which must consider not only the energy constraint in sensor nodes, but also the processing and storage capacities of sensor nodes.

IV. Processing Power: Sensor nodes will have a processing power in the order of a few MIPS. This takes more processor time if the task is running for long time and preventing other jobs to wait for longer time irrespective of their priorities. Hence, operating system should properly schedule the processor according to the priority of jobs.

V. Massive and Random Deployment: Most sensor networks consist of a large number of sensor nodes, from hundreds to thousands or even more. Node deployment is usually application dependent, which can be either manual or random. In most applications, sensor nodes can be scattered randomly in an intended area or dropped massively over an inaccessible or hostile region. The sensor nodes must autonomously organize them-selves into a communication network before they start to perform a sensing task.

VI. Memory: The current generation of micro-controllers family such as Mica, its successors and some microcontrollers specific to various research projects have nearly 128kbytes of program memory. The system software such as operating system, virtual machine, middleware, and application algorithms have to fit into this memory. Sensor nodes have non volatile external data storage mechanism EEPROM or flash memory.

VII. Dynamic and Unreliable Environment: A sensor network usually operates in a dynamic and unreliable environment. On one hand, the topology of a sensor network may change frequently due to node failures, damages, additions, or energy depletion. On the other hand, sensor nodes are linked by a wireless medium, which is noisy, error prone, and time varying. The connectivity of the network may be frequently disrupted, because of channel fading or signal attenuation.

VIII. Bandwidth: A typical sensor node uses RF channel to communicate with other sensor nodes in the network. ZigBee is the emerging standard to define the communication protocol stack based on the existing physical and data-link layers of IEEE 802.15.4 Personal Area Network (PAN) standard.

Data rate supported by PANs is 256kbps. Whereas Bluetooth standard supports data rate up to 3Mbps. CC1000 is another standard that has been widely used in sensor networks. Its data rate is around 39kbps. Rarely used wireless standard in WSN is IEEE 802.11 (Wi-Fi), whose data rate is almost 54Mbps. Hence, the bandwidth constraint is the basic factor for the dynamic network scenario.

IX. Diverse Applications: Sensor networks have a wide range of diverse applications. The requirements for different applications may vary significantly. No network protocol can meet the requirements of all applications. The design of sensor networks is application specific.

3.14 CROSS LAYER APPROACH

Cross layer design may be defined as, "the contravention of OSI hierarchical layers in communication networks" or "protocol design by the violation of reference layered communication architecture is cross-layer design with respect to the particular layered architecture". The breaking of OSI hierarchical layers or the violation of reference architecture as shown in Fig. 3.8 includes merging of layers, creation of new interfaces, or providing additional interdependencies between any two layers.

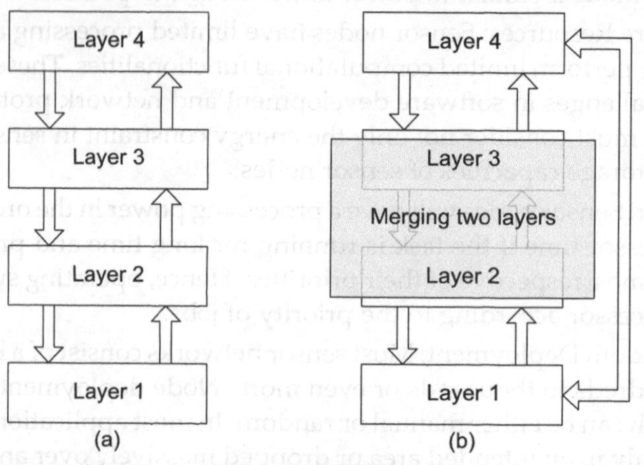

Fig. 3.8 Layer architecture of interfaces (Fig. (a)) and its cross layer (Fig. (b))

There are several challenges those tradeoffs and optimizations parameters of the network scale and the system throughput, the power consumption and the system lifetime. The solution is Cross-layer protocol interactions that can lead network efficiency and better QoS support. A Cross-Layer Design (CLD) is particularly important for any network using wireless technologies, since the information exchange between different layers can optimize the network throughput.

3.15 CROSS LAYER PROTOCOL DESIGN

Cross-Layer has recently become the new concept in wireless communication systems. Its Design has a great potential in future wireless communication system. Cross-layer based MAC protocol is to improve the channel bandwidth and to increase the fairness of each flow without causing congestion. The Fig. 3.9 shows standardized cross layer interface between the different layers, and thus able to decouple the individual protocol layers. This allows for dynamic insertion communication between the different layers to "transmit" and "receive" signal where "transmit" is invoked by the upper layer to transfer data to the lower layer. Equally "handle receive" is invoked by the lower layer to transfer data to the upper layer.

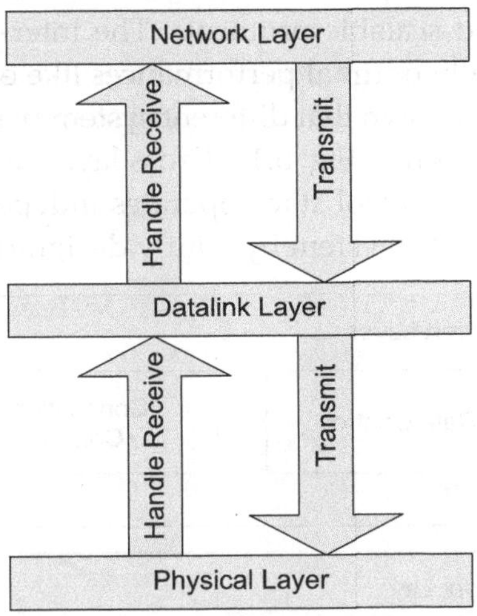

Fig. 3.9 Structure of a three layer protocol stack

Protocol stacks has a huge limitation, due to the dynamic applications. An application detects the Physical layer force to use a different medium for its data transport. The application is then given the option to change the lower layers of the protocol, while keeping the upper layers intact.

To increase the lifetime of the resource constraints network in WSN, node requires energy efficient and energy aware schemes on all layers of the protocol stack. The traditional layered networking approach has several drawbacks from WSNs perspective, improvements in performance and energy efficiency are possible if significant amount of information is passed across protocol layers and hence network lifetime can be improved.

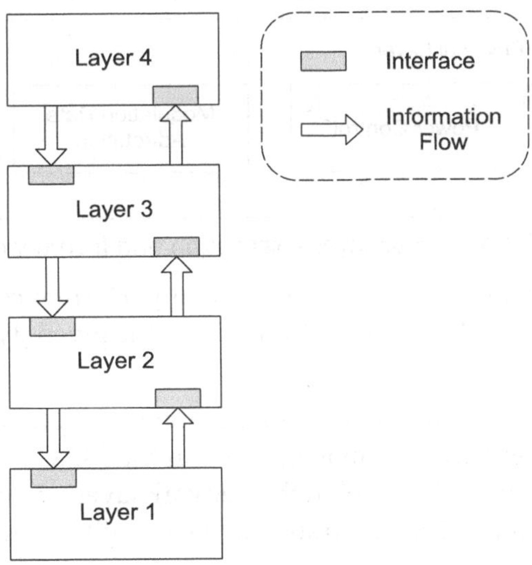

Fig. 3.10 Illustrative reference architecture

Here, each layer has defined interfaces for communication with any other layers. For example, in Fig. 3.10, layer 3 can only communicate with adjacent layers (layer 4 and layer 2) via defined interfaces and it cannot communicate with layer 1 as no such interface is available for information exchange.

The simplicity of design of a layered protocol stack interfaces between independent layers result in the development of robust and scalable protocols. The inter-dependencies between different layers can utilized to get statistically optimal performances like energy efficiency due to enhancement of the cross layer design. It is known that different system parameters are controlled in distinct layers in a wireless network as shown in Fig. 3.11. Cross layer design is different from traditional network design where each layer protocol stack operates independently. A cross layer approach seeks to enhance the performance of a system by jointly designing multiple protocol layers.

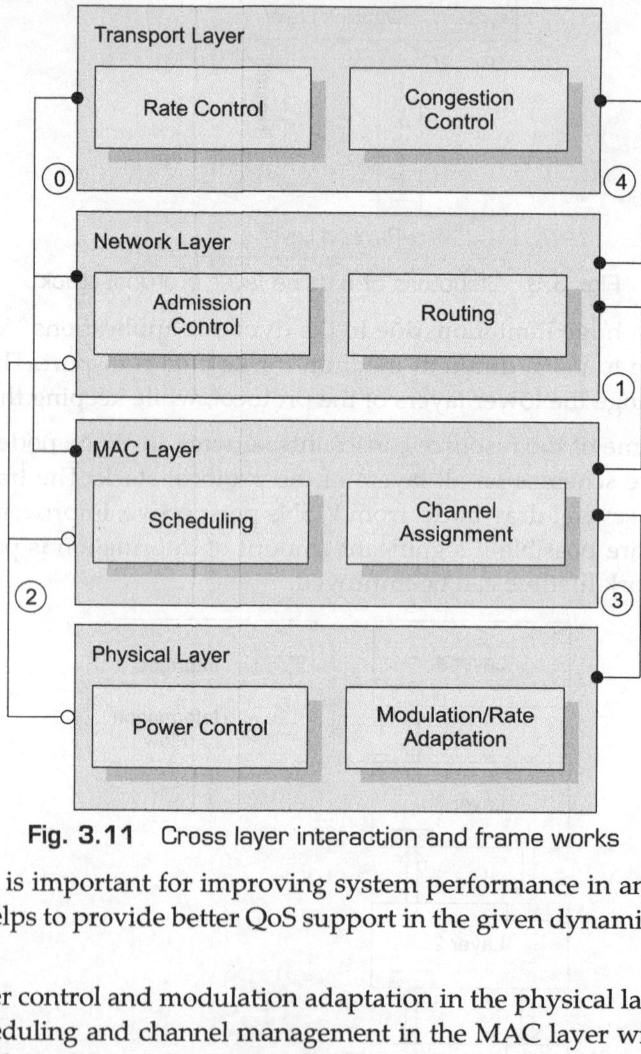

Fig. 3.11 Cross layer interaction and frame works

Cross-layer design is important for improving system performance in an ad-hoc network. The resulting flexibility helps to provide better QoS support in the given dynamic network and limited resources.

For example, power control and modulation adaptation in the physical layer change the overall system topology. Scheduling and channel management in the MAC layer will affect the space and time reuse in a network. Routing and control in the network layer will change the flow distribution. Finally, congestion and rate control in the transport layer will change the traffic volume in the communication link.

In recent years, there are researches using cross-layer design for high efficiency and low cost multi-hop wireless communication systems. These efforts can be met by cross-layer design to improve the power efficiency, improving system throughput in wireless networks. Multiple reasons to motivate the cross-layer research for the wireless sensor networks are taking place. The goal is to provide a feasible and flexible approach to solve the conflicts between the requirements of large scale, long life-time, and multiple purpose wireless sensor networks and the constraints of tight bandwidth, low battery capacity, and limited node resources. Hence there is need to optimize the cross layer approach. As shown in Table 3.1, the optimizing approach at each layer. It shows that some solutions for those three optimization goals are either conflict with each other or orthogonal to each other. The network layer and transport layer that handle the end to end data transmissions will be of no use in this application.

Table 3.1 Optimizing approach of each layer.

Layer	Network Scale	System Life-time	Node versatility
Application	Data fusion, Compression	Power-aware mode control	Load detection, Automatic mode decision
Transport	Bounded Delay	QoS-power tradeoff	Load-aware transport control
Network	Node naming, Efficient routing, Efficient node discovery	Power-aware routing, Reduced overhead	Load-aware routing, Simplified node discovery, Distributed storage
MAC	Contention control, Channel reuse	Synchronized sleep, Transmission range control	Load-aware channel allocation
Physical	Ultra-wide Band	Low-power design, Powerful battery	Attach specific accessories (GPS)

I. Optimization can be achieved in multiple layers. All three optimization goals are targeting at can be achieved in all five layers of the system. This fact provides a rationale for the cross-layer optimizations to achieve our research goal.

II. Optimization in one layer depends on the other layers to show its effects. Cross-layer optimization is necessary because it is possible that different approaches for the same optimization target may counteract each other. For power saving there is need to design a power-saving routing protocol to select the shortest routes so that they will pass the most densely deployed area. This kind of routes may take advantage of the fact that for the same transmission distance more number of smaller-distance hops will save transmission powers compared with a single larger-distance hop.

If the MAC layer is not optimized due to contention, the advantages of the routing design may be counteracted by the increasing power consumption due to the increase of contention possibility.

The goal of the cross layer is to provide a feasible and flexible approach to solve the conflicts between the requirements of large scale, long life-time, and multiple purpose wireless sensor networks and the constraints of tight bandwidth, low battery capacity, and limited node resources. The cross-layer optimization is a promising solution for optimization.

3.16 CLUSTERING IN WSN

In recent years, Wireless Sensor Network (WSN) has attracted much interest in the wireless research community as a fundamentally new tool for a wide range of monitoring and data-gathering applications. There have been major advances in the development of low power micro sensor nodes that led practitioners to envision networking a large set of sensors scattered over a wide area of interest into a wireless sensor networks (WSN) for large-scale event monitoring and data collection and filtering. Sensor nodes are significantly constrained in the amount of available resources such as energy, storage, and computational capacity. Due to energy constraints, a sensor can communicate directly only with other sensors that are within a small distance. To enable communication between sensors not within each other's communication range, sensors form a multi-hop communication network.

One of the major and probably most important challenges in the design of WSNs is the fact that all nodes will have to rely on a limited supply of energy. Replacing these resources in the field may be difficult or impossible, causing severe limitations in the communication and processing time between all sensors in the network. Clustering is particularly practical for application that requires scalability to hundreds or thousands of nodes.

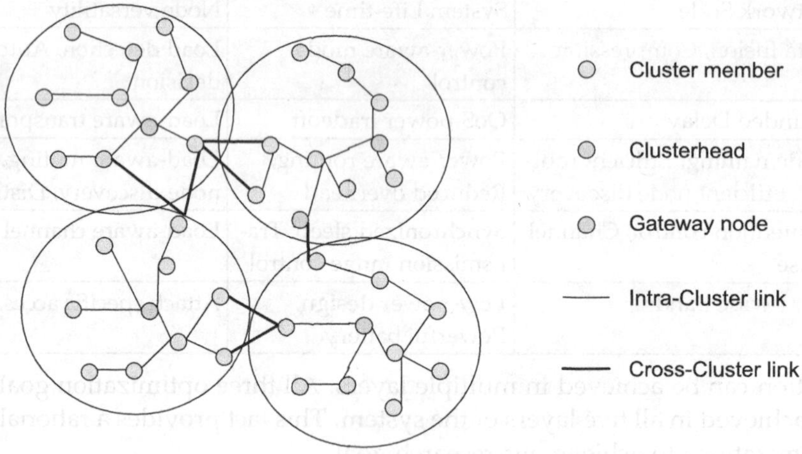

Fig. 3.12 Cluster nodes distributions

Clustering is an effective topology control approach and its distribution in wireless sensor networks is shown in Fig. 3.12. The essential operation in sensor node clustering is to select a set of cluster heads from the set of nodes in the network, and then cluster the remaining nodes with these heads. Cluster heads coordinate among the nodes within their clusters and aggregation of their data (intra-cluster coordination), and communication with each other and/or with external observers on behalf of their clusters (inter-cluster communication).

Clustering is a standard approach for achieving efficient and scalable performance in sensor networks. Clustering facilitates the distribution of control over the network and, hence, enables locality of communication. Moreover, clustering nodes into groups saves energy and reduces network contention as nodes communicate their data over shorter distances to their respective cluster-heads.

Many protocols have been proposed for sensor networks in the last few years. Reducing energy consumption has been primarily addressed. Clustering is a standard approach for achieving efficient and scalable performance in wireless sensor networks. Traditionally, clustering algorithms aim at generating a number of disjoint clusters that satisfy some criteria.

Wireless sensor network is a new network paradigm that involve the deployment of hundreds-even thousand-of low-cost, energy-limited, small, and application specific sensor nodes to create applications for factory monitoring and control, disaster response, military sensing intelligent house control, and, etc. Since, WSN is typically deployed in an uncontrolled or unreachable environment, each sensor node carries a limited, generally irreplaceable energy source. Therefore, energy conservation is the most important performance objective to extend network lifetime while designing WSN protocols, such as media access control, routing data, aggregation for WSN.

3.17 CLUSTER ARCHITECTURE

Clustering is defined as the grouping of similar objects or the process of finding a natural association among some specific objects or data. Clustering is useful for application that requires scalability to hundreds or thousands of nodes. Scalability in this context implies the need for load balancing, efficient resource utilization, and data aggregation. Formation of clusters is to transmit processed data to base stations, hence minimizing the number of nodes that take part in long distance communication

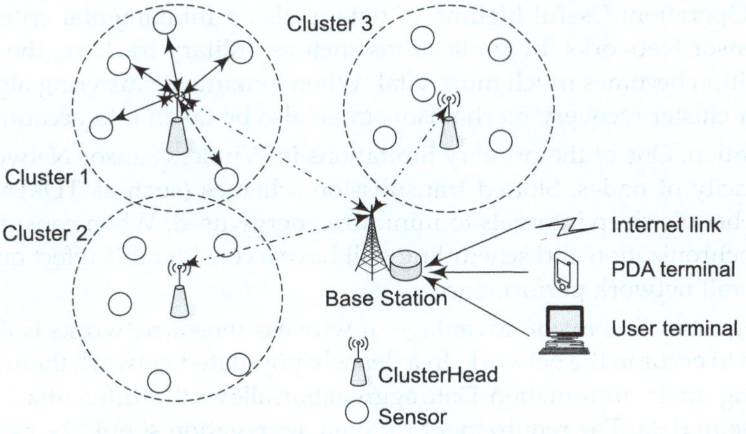

Fig. 3.13 General sensor network architecture

As illustrated in the Fig. 3.13, the architecture of the clustering of the organizational structure in the Wireless Sensor Network. Following are some of the areas:

(*i*) Sensor Node: A sensor node is the core component of a WSN. It take on multiple roles in a network, such as simple sensing; data storage; routing; and data processing.

(*ii*) Clusters: Clusters are the organizational unit for WSNs. The dense nature of these networks had broken down into clusters to simplify tasks such a communication.

(*iii*) Cluster heads: Cluster heads are the organization leader of a cluster. They often are required to organize activities in the cluster. Their tasks limited to data-aggregation and organization the communication schedule of a cluster.

(*iv*) Base Station: The base station is at the upper level of the hierarchical WSN. It provides the communication link between the sensor network and the end-user.

(*v*) End User: This query is generated by the end user. In a queried sensor network the required data is gathered from a query sent through the network such as use of the network data over the internet, using a PDA, or even a desktop computer. .

The clustering phenomenon plays an important role not just organization of the network, but can dramatically affect network performance.

3.18 CLUSTER DESIGN PHILOSOPHY

Wireless Sensor Networks present vast challenges in terms of implementation. Design goals targeted in traditional networking provide little more than a basis for the design in wireless sensor networks. Clustering algorithms play a vital role in achieving the targeted design goals for a given implementation. There are several key attributes that designers must carefully consider, which are of particular importance in wireless sensor networks.

(*a*) Cost of Clustering: Although, clustering plays a vital role in organizing sensor network topology, there are often many resources such as communication and processing tasks needed in the creation and maintenance of the clustering topology. Such costs as the required resources are not being used for data transmission or sensing tasks.

(*b*) Selection of Cluster heads and Clusters: The clustering concept offers tremendous benefits for wireless sensor networks. However when designing for a particular application, designers must carefully examine the formation of clusters in the network depending on the application.

(*c*) Real-Time Operation: Useful lifetime of data is also a fundamental criterion in designing Wireless Sensor Networks. In applications such as military tracking, the issue of real-time data acquisition becomes much more vital. When looking at clustering algorithms, the time required for cluster recovery mechanisms must also be taken into account.

(*d*) Synchronization: One of the primary limitations in Wireless Sensor Networks is the limited energy capacity of nodes. Slotted transmission schemes (such as TDMA), allow nodes to regularly schedule sleep intervals to minimize energy used. When considering a clustering scheme, synchronization and scheduling will have a considerable effect on network lifetime and the overall network performance.

(*e*) Data Aggregation: One major advantage of wireless sensor networks is the ability for data aggregation to occur in the network. In a densely populated network there are often multiple nodes sensing similar information. Data aggregation allows the differentiation between sensed data and useful data. The requirement for data aggregation should be carefully considered when selecting a clustering approach.

(*f*) Repair Mechanisms: Due to the nature of Wireless Sensor Networks, they are often prone to node mobility, node death and interference. All of these situations can result in link failure. When looking at clustering schemes, it is important to look at the mechanisms in place for link recovery and reliable data communication.

(*g*) Quality of Service (QoS): From an overall network stand- point, we can look at QoS requirements in Wireless Sensor Networks. Many of these requirements are application dependant and as such, it is important to look at these metrics when choosing a clustering scheme. Implementations can vary widely in terms of these metrics, and as a result, the design process should consider these aspects.

3.19 LIMITATION OF CLUSTERING

There are several key limitations in WSNs, that clustering schemes must consider.

(*i*) Limited Energy: Unlike wired designs, wireless sensor nodes are "off-grid", meaning that they have limited energy storage and the efficient use of this energy will be vital in determining the range of suitable applications for these networks. The limited energy in sensor nodes must be considered as proper clustering can reduce the overall energy usage in a network.

(*ii*) Network Lifetime: The energy limitation on nodes results in a limited network lifetime for nodes in a network. So, proper clustering should attempt to reduce the energy usage, and hereby increase network lifetime.

(*iii*) Limited Abilities: The small physical size and small amount of stored energy in a sensor node limits many of the abilities of nodes in terms of processing and communication abilities. A good clustering algorithm shall make use of shared resources within an organizational structure, while taking into account the limitation on individual node abilities.

(*iv*) Application Dependency: Often a given application will heavily rely on cluster organization. When designing a clustering algorithm, application robustness must be considered as a good clustering algorithm should be able to adapt to a variety of application requirements.

3.20 CLUSTER FORMATION ALGORITHM

In Cluster formation routing algorithm each node transmits some packets named "Hello message" to announce its presence to its neighbor nodes. Upon receiving a hello message, each node updates its neighbor tables. Each node enters the network in the "undecided" state. Every node upon receiving hello message from its neighbors compares its own ID with its neighbor's. If a node distinguishes that its own ID is the lowest ID between its neighbors, this node declares itself as cluster head. Every node that has a bi-directional link to this cluster head will be a member of this cluster.

Clusters are identified by their respective cluster heads, which means that the cluster head must change as infrequently as possible. In Fig. 3.14 shown, node 1 is cluster head for the cluster containing nodes 2, 3, 4 and 5, and for other two nodes 6 and 8 they are cluster heads.

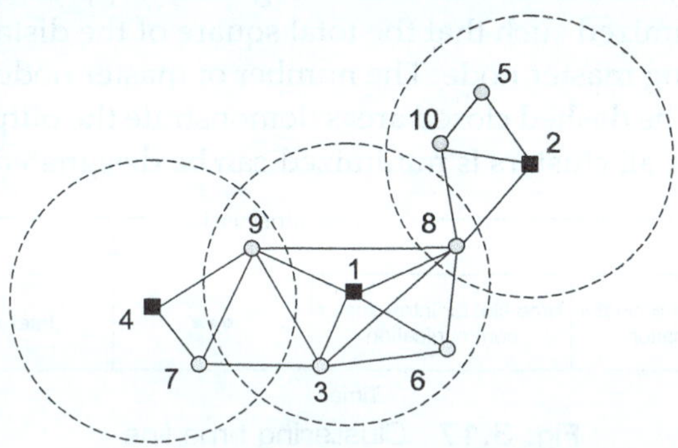

Fig. 3.14 Cluster structure routing

Clustering algorithm creates clusters based on selected diameters. Here, the essential operation in sensor node clustering is to select a set of cluster heads from the set of nodes in the network, and then cluster the remaining nodes with these heads. Cluster heads are responsible for coordination among the nodes within their clusters and aggregate their data (intra-cluster coordination), and communication with each other and/or with external observers on behalf of their clusters (inter-cluster communication).

- Intra-cluster routes (routes within a cluster) are discovered dynamically using the membership information.
- Inter-cluster routes (routes between clusters) are found by flooding the network with Route Requests (RREQ).

- Routing is based on source routing and the route discovery is done by flooding the network with RREQ. RREQ will always follow a route with the following pattern:
  [Source → Cluster head → Gateway → Cluster head → Gateway → ··· → Destination]

Here, below Fig. 3.15, shows sample sensor network with 24 sensor nodes. Here, the same sensor network with a different capacity constraint for master nodes. Each master node can handle 4 sensors in this example. Therefore, the number of master nodes has been increased in order to cover all sensor nodes.

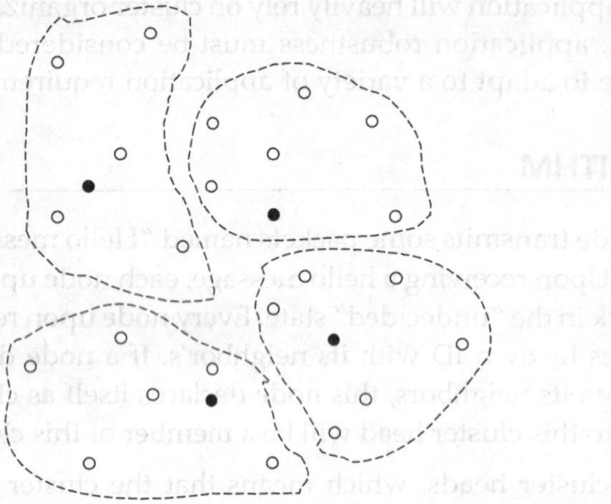

Fig. 3.15 Each MN handles 6 sensors

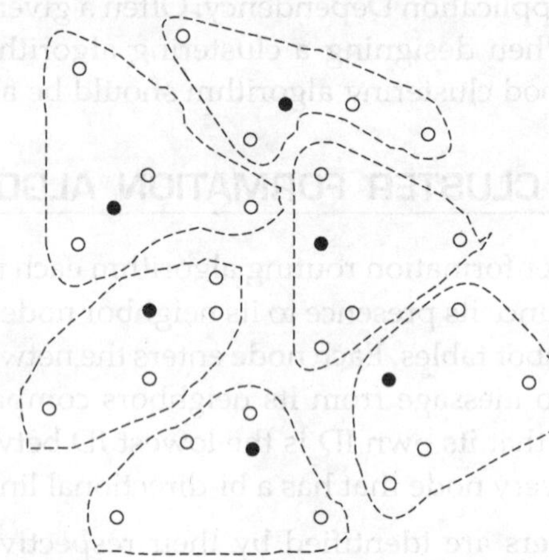

Fig. 3.16 Each MN can handle 4 sensors

The sensors are randomly distributed on a small region. By applying algorithm to form k clusters, over all clusters is minimized such that the total square of the distances between sensors in a cluster and the corresponding master node. The number of master nodes depends on the capacity constraint of each of them. The dashed closed areas demonstrate the output of algorithm, hence the optimal solution where over all clusters is minimized can be designated as MN=?.

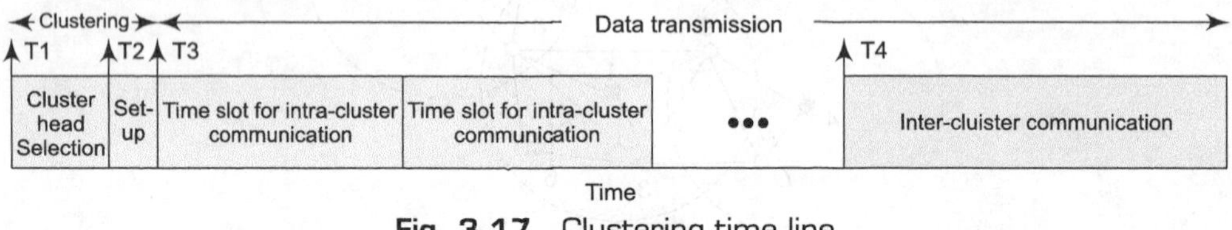

Fig. 3.17 Clustering time line

The Fig. 3.17 illustrates the operation of unequal cluster routing (UCR) time line protocol for one data gathering round. After each cluster head has chosen a relay node decided to transmit its data to the base station directly, a tree rooted at the base station is constructed. A cluster head receives data packets from tree descendants and sends them with the cluster's own packets up to the root.

It begins with a clustering phase when cluster heads are selected and the intra-cluster TDMA schedule are set-up, followed by a data transmission phase, where data are transferred from the nodes to the cluster head and on to the base station via a multihop path.

In a WSN, there are several clusters based protocols such as LEACH, TEEN, PEGASIS, EHIP, EECP have been proposed. However, these algorithms some consider clustering, node delivers sensed data to data sink through its cluster-head.

SUMMARY

This chapter illustrates an energy-efficient routing protocol based on clustering and least spanning tree for wireless sensor network prolong network lifetime and shorten path. Clustering includes partitioning stage and choosing stage, namely, partitions the multi-hop network and then choose, cluster head for receiving, sending and maintaining information in its cluster.

Chapter explains the architecture of cross layer and clusters. Optimal clustering in terms of energy efficiency should eliminate all overhead associated not only with the cluster head selection process, but also with node association to their respective cluster heads.

QUESTIONS

1. What are the different issues in WSN?
2. Explain the layer protocol.
3. Explain, briefly the issues of data link layers.
4. Explain, briefly the issues of network layers.
5. Explain, briefly the issues of transport layers.
6. What is meant by cluster and its importance?
7. Draw out the network area classification.
8. What are the challenges in network design?
9. What is cross layer? Explain its advantages over single layer.
10. Draw and explain the cluster architecture.
11. Explain the philosophy of cluster design.
12. Explain the cluster algorithm.

BIBLIOGRAPHY

- Agre, J.R. et al. Development platform for self-organizing wireless sensor networks, SPIE – The International Society for Optical Engineering, 3713, 257, 1999.
- Culler, D.; Estrin, D.; Srivastava, M. Overview of sensor networks. IEEE Comput. Mag. 2004, 37, 41–49.
- Rajaravivarma, V.; Yang, Y.; Yang, T. An Overview of Wireless Sensor Network and Applications. In Proceedings of 35th Southeastern Symposium on System Theory, Morgantown, WV, USA, 2003; pp. 432–436.
- Akyildiz, I.; Su, W.; Sankarasubramaniam, Y.; Cayirci, E. A survey on sensor networks. IEEE Commun. Mag. 2002, 40, 102–114.
- Hill, J. et al. System architecture directions for networked sensors, ASPLOS, 2000, 93.
- Min, R. et al. An architecture for a power-aware distributed microsensor node, in Proc. IEEE Workshop on Singal Processing Systems, 2000, 581.
- Ye, W., Heidemann, J., and Estrin, D. An energy-efficient MAC protocol for wireless sensor networks, in Proc. INFOCOM, 3, 1567, 2002.
- K. R¨omer and F. Mattern. The Design Space of Wireless Sensor Networks IEEE Wireless Communications 11(6), December 2004.
- J. Polastre, R. Szewczyk and David Culler. Telos: Enabling Ultra-Low Power Wireless Research IEEE IPSN, April 2005.
- D. Gay, P. Levis, R. von Behren, M. Welsh, E. Brewer and D. Culler. The nesC Language: A Holistic Approach to Networked Embedded Systems ACM PLDI, June 2003.

- Dunkels. Full TCP/IP for 8-Bit Architectures ACM MOBISYS 2003, May 2003.
- Mhatre, C. Rosenberg, "Design Guidelines for Wireless Sensor Networks: Communication, Clustering and Aggregation," Ad-Hoc Networks Journal, Elsevier Science, 2004.
- Y. Ganjali, A. Keshavarzian, "Load Balancing in Ad-Hoc Networks: Single -path Routing vs. Multi-path Routing," IEEE Infocom'2004.
- Intanagonwiwat, R. Govindan, D. Estfin, J. Heidemann, and F. Silva. "Directed diffusion for wireless sensor networking". IEEE/ACM Transactions On Networking, 1(1), February 2003.
- S. Lindsey, C. Raghavendra, and K. M. Sivalingam. "Data gathering algorithms in sensor networks using energy metrics". IEEE Transactions on Parallel and Distributed Systems, 2002.
- M. Ulema, J. M. Nogueira, and B. Kozbe. Management of wireless ad hoc networks and wireless sensor networks. Journal of Network and Systems Management, 14(3): 327–333, September 2006.
- C., R. Govindan, and D. Estrin, Directed Diffusion: A Scalable and Robust Communication Paradigm for Sensor Networks. 2000: Mobile Computing and Networking.
- Culler, D.E. et al. Network-centric approach to embedded software for tiny devices, in Proc. Workshop on Embedded Software, 2001, 114.

4

MAC Protocol in WSN

4.1 INTRODUCTION

Wireless sensor networks (WSNs) constitute a special class of wireless data communication networks with large number of nodes equipped with embedded processor, sensor, drivers and radios. It is a group of specialized transducers with a communications infrastructure intended to monitor and record conditions at diverse locations.

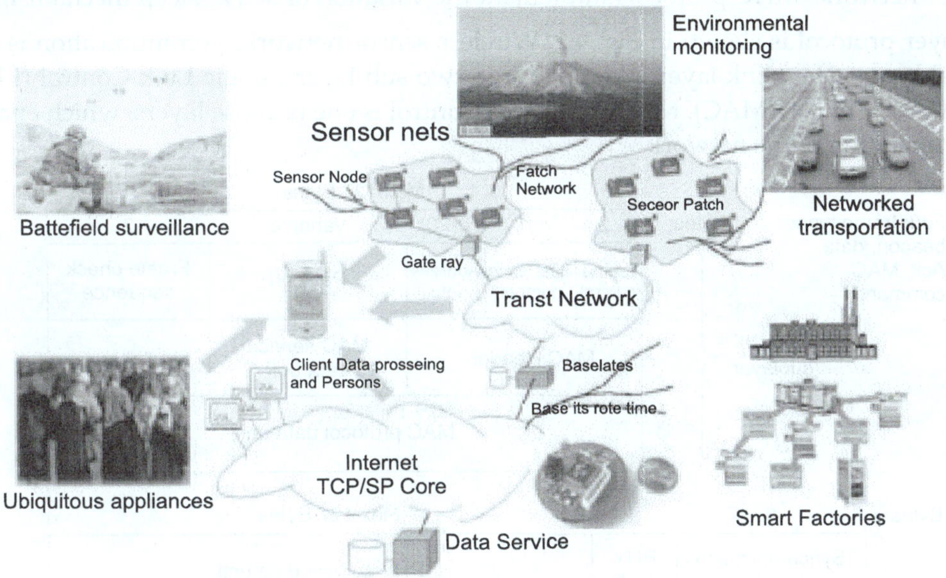

Fig. 4.1 Wireless sensor network areas

A WSN consists of multiple detection stations called sensor nodes, each of which is small, lightweight and portable as shown in Fig. 4.1. Every sensor node is equipped with a transducer, microcomputer, transceiver and power source. The transducer generates electrical signals based on sensed physical effects and phenomena. The microcomputer processes and stores the sensor out-put.

The transceiver receives commands from a central computer and transmits data to that computer. The power for each sensor node is derived from the electric utility or from a battery.

The primary objective of WSN is to maximize the sensor node/network lifetime. Sensor nodes are generally battery powered. The improvements in hardware technology have resulted in low-cost sensor nodes. This sensor node is more energy-consuming than their computation. Low power capacities have lead to limited coverage and communication range for sensor nodes when compared to other mobile devices. Hence, it is an emerging paradigm posing new challenges for researchers in wireless communications.

It is generally hard (or impractical) to charge/replace the exhausted battery in the WSN. So, the medium access decision within a dense network composed of nodes with low duty-cycles is a hard problem. It needs to be solved in an energy-efficient manner to maximize the node/network life-time, leaving the other performance metrics as secondary objectives.

Communication in wireless sensor networks is divided into several layers. Medium Access Control (MAC) layers protocol tries to avoid collisions by not allowing two interfering nodes to transmit at the same time. This will enables for the successful operation of the network.

The main design goal of a typical MAC protocols is to provide high throughput and QoS. On the other hand, wireless sensor MAC protocol gives higher priority to minimize energy consumption than QoS requirements. Energy gets wasted in traditional MAC layer protocols due to idle listening, collision, protocol overhead, and overhearing.

There are some MAC protocols that have been especially developed for wireless sensor networks. Typical examples include S- MAC, T-MAC, and D-MAC etc. To maximize the battery life-time, sensor networks MAC protocols implement the variation of active/sleep mechanism.

MAC layer protocol is shown in Fig. 4.2. Wireless sensor networks communication is divided into several layers. Data link layer is divided into two sub layers: Logic Link Control (LLC) and Medium Access Control (MAC). Medium Access Control is one of those layers, which enables the successful operation of the network.

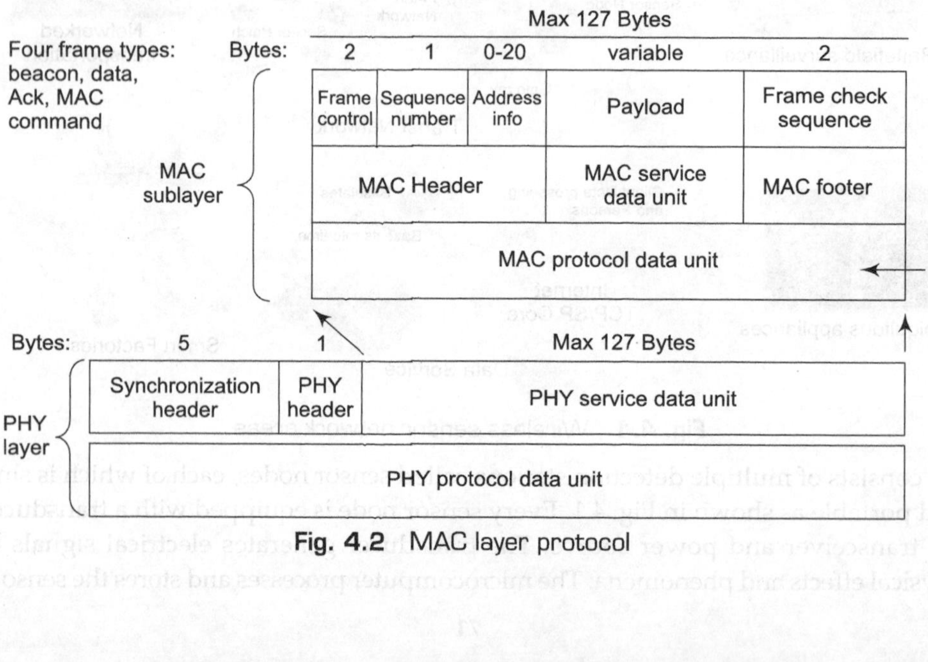

Fig. 4.2 MAC layer protocol

MAC protocol tries to avoid collisions by not allowing two interfering nodes to transmit at the same time. The main design goal of a typical MAC protocols is to provide high throughput and QoS. On the other hand, wireless sensor MAC protocol gives higher priority to minimize energy consumption than QoS requirements. Energy gets wasted in traditional MAC layer protocols due to idle listening, collision, protocol overhead, and overhearing.

4.2 MAC PROTOCOL OF 802.11

The various types of multiplexing used in 802.11 are TDMA (Time Division Multiple Access), FDMA (Frequency Division Multiple Access), CDMA (Code Division Multiple Access) and STDMA (Space and Time Division Multiple Access). On a different level 802.11 multi-accesses can be classified into centralized and distributed in the Table 4.1.

Table 4.1 Difference between centralized and distributed topology

Centralized	Distributed
1. Management is easy.	1. Management is difficult.
2. Service differentiation is possible.	2. Service differentiation is difficult.
3. Single point of failure.	3. No single point of failure.
4. Poor scalability.	4. Scalability is easy.

BSS (Basic service set): A BSS is defined as a set of radios talking to each other in the vicinity of each other. A BSS can be classified as an IBSS (Independent BSS) or and Infrastructure BSS. In IBSS peer-to-peer communication takes place in Ad-hoc mode. In infrastructure BSS there exists one access point and all stations communicate to this access point as shown in Fig. 4.3.

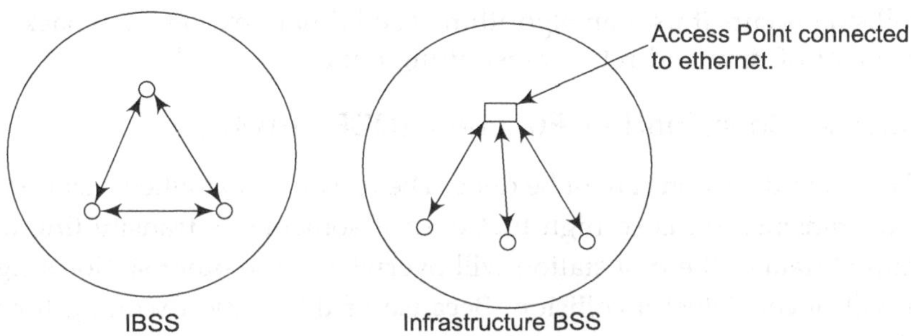

Fig. 4.3 BSS and IBSS

In infrastructure BSS the access points are connected to the Ethernet. However, another arrangement has been defined, called ESS (Extended Service Set), in which access points themselves may communicate with each other through the wireless network. The network in this situation is called Wireless Distribution Network. When a user moves from coverage area of one access point to the coverage area of second access point it is taken care of by the access points and it is transparent to the user.

There are two MAC protocols defined for 802.11: Centralized and Distributed as shown in Fig. 4.4. The centralized protocol is called Point Coordination Function (PCF) and the distributed protocol is called Distributed Coordination Function (DCF). It is interesting to note that industry equipment usually has only distributed MAC protocol implemented. The DCF protocol is quite similar to the

Ethernet MAC protocol CSMA/CD (Carrier Sense Multiple Access/Collision Detect). Hence, salient features of CSMA/CD are in order.

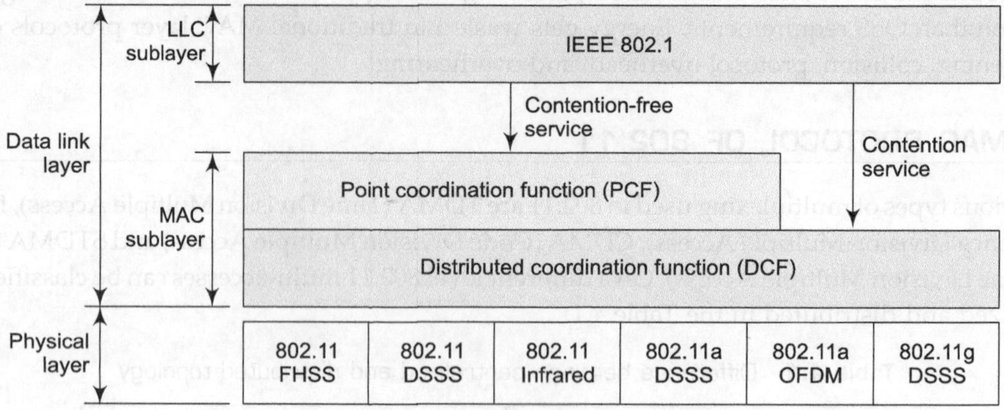

Fig. 4.4 MAC layer in 802.11 standards

4.2.1 Salient Features of CSMA/CD

(*i*) Each station first listens to the channel before transmitting. If the channel is busy it will wait for channel become free. (Carrier Sense)

(*ii*) As soon as a station wanting to transmit data finds the channel free it transmits the data with probability 1(called 1-persistent).

(*iii*) While transmitting a station also listens to the channel and compares the data that it has transmitted with the data it has received. If the data are different a collision has occurred and the channel discontinues transmitting. (Collision detect)

(*iv*) When a collision occurs it uses an algorithm called Binary exponential back off to back off a random amount of time and retry transmitting data.

4.2.2 Distributed Coordination Function (DCF) Protocol

In DCF protocol collision detection cannot be done. The reason is that when a station is transmitting data the power of transmission is so high that even if someone is transmitting at that time the power of transmitted data of the first station will overwhelm the other station's signal and hence the first station will never detect a collision. Because of this station waiting for the channel to become free and cannot start transmitting as soon as they sense that the channel is free, because if they do so then the probability of collision occurring is very high and they will not even be able to detect it. Hence, DCF protocol uses following rules:

(*i*) Listen before transmission. (Carrier Sense)

(*ii*) Backoff before transmission.

(*iii*) Backoff on collision.

Consider the scenario as shown the DIFS (DCF Inter-Frame Space) of A, B and C in Figs. 4.5 and 4.6.

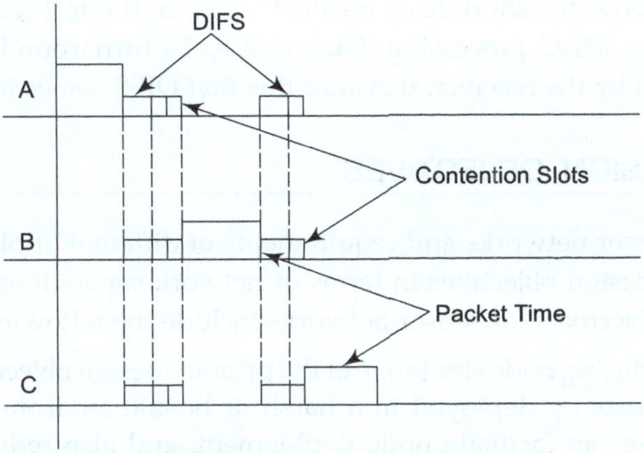

Fig. 4.5 DIFS time frame of A, B, and C

'A' is transmitting its frame in a small time called DIFS. The 'A', 'B' and 'C' select randomly a number between 0 and their respective contention window variable 'cw'. Let's us assume 'A' selected 5, 'B' selected 2 and 'C' selected 3. Then as shown all three stations wait for DIFS time and then for 2 contention slot time. While waiting for each contention slot time all stations reduce their counters by 1 for each slot time waited. Thus after waiting for 2 slots counter of 'B' becomes 0 and hence it starts transmitting. When any channel starts transmitting all other waiting stations stop decrementing their counters. After 'B' has completed transmission again all stations wait for DIFS time and then start reducing their counters. When 'B' started transmitting the counter value of 'C' was 1.

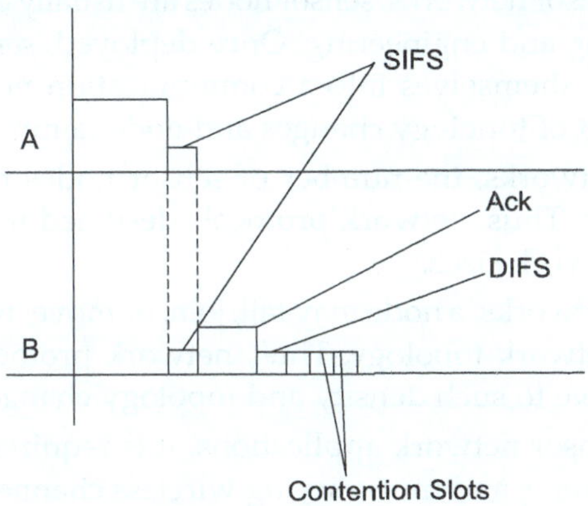

Fig. 4.6 DIFS time frame of A and B

Thus, after 'B' completed transmission 'C' waited for one slot and then started transmitting. After transmission is completed the transmitting station again chooses randomly a number between 0 and 'cw' if it wants to transmit data.

- Minimum value of 'cw' is 32 and its
- Maximum value of 'cw' is 1024.

Initial value of 'cw' is 32. Whenever a collision occurs, 'cw' of the colliding stations is doubled. After this the colliding stations again choose randomly a number between 0 and 'cw'. On successful transmission value of 'cw' is reset to 32. Standard contention slot time is 20 microseconds.

The diagram shown above the short delay is called SIFS (Short InterFrame Space). SIFS includes RxRF Delay, PLCP Delay, MAC processing delay and RxTx turnaround delay. After this delay acknowledgement is sent by the receiver. It is after this that DIFS comes into picture.

4.3 NETWORK DESIGN OBJECTIVES

The characteristics of sensor networks and requirements of different applications have a decisive impact on the network design objectives in terms of network capabilities and network performance. The main design objectives for sensor networks include the following several aspects:

- *Small Node Size.* Reducing node size is one of the primary design objectives of sensor networks. Sensor nodes are usually deployed in a harsh or hostile environment in large numbers. Reducing node size can facilitate node deployment, and also reduce the cost and power consumption of sensor nodes.
- *Low Node Cost.* Reducing node cost is another primary design objective of sensor networks. Since, sensor nodes are usually deployed in a harsh or hostile environment in large numbers and cannot be reused, it is important to reduce the cost of sensor nodes so that the cost of the whole network is reduced.
- *Low Power Consumption.* Reducing power consumption is the most important objective in the design of a sensor network. Since, sensor nodes are powered by battery and it is often very difficult or even impossible to change or recharge their batteries. So, it is crucial to reduce the power consumption of sensor nodes so that the lifetime of the sensor nodes, as well as the whole network is prolonged.
- *Self-Configurability.* In sensor networks, sensor nodes are usually deployed in a region of interest without careful planning and engineering. Once deployed, sensor nodes should be able to autonomously organize themselves into a communication network and reconfigure their connectivity in the event of topology changes and node failures.
- *Scalability.* In sensor networks, the number of sensor nodes may be on the order of tens, hundreds, or thousands. Thus, network protocols designed for sensor networks should be scalable to different network sizes.
- *Adaptability.* In sensor networks, a node may fail, join, or move, which would result in changes in node density and network topology. Thus, network protocols designed for sensor networks should be adaptive to such density and topology changes.
- *Reliability.* For many sensor network applications, it is required that data be reliably delivered over noisy, error-prone, and time-varying wireless channels. To meet this requirement, network protocols designed for sensor networks must provide error control and correction mechanisms to ensure reliable data delivery.
- *Fault Tolerance.* Sensor nodes are prone to failures due to harsh deployment environments and unattended operations. Thus, sensor nodes should be fault tolerant and have the abilities of self-testing, self-calibrating, self- repairing, and self-recovering.
- *Security.* In many military applications, sensor nodes are deployed in a hostile environment and thus are vulnerable to adversaries. In such situations, a sensor network should introduce effective security mechanisms to prevent the data information in the network or a sensor node from unauthorized access or malicious attacks.

- *Channel Utilization.* Sensor networks have limited bandwidth resources. Thus, communication protocols designed for sensor networks should efficiently make use of the bandwidth to improve channel utilization.
- *QoS Support.* In sensor networks, different applications may have different quality-of-service (QoS) requirements in terms of delivery latency and packet loss. For example, some applications, for example, fire monitoring, are delay sensitive and thus require timely data delivery. Thus, network protocol design should consider the QoS requirements of specific applications.

4.4 MAC PROTOCOL DESIGN ISSUE

Wireless Ad-hoc networks are comprised of dynamic nodes that exchange packets by sharing a common broadcast radio channel. Due to the limitations of the channel, the bandwidth needs to be shared among the nodes. Therefore, the networks goal is to utilize the bandwidth efficiently, and guarantee fairness to all nodes. The ad-hoc wireless networks have more specific characteristics, such as node mobility, power constraints thus new protocols are needed for controlling access to the physical medium. The unique properties of the ad-hoc networks make the design of a media access control (MAC) protocol more challenging.

Following are the main issues one should have in mind when considering designing a MAC protocol for ad-hoc wireless networks.

- *Bandwidth efficiency:* The scarcity of bandwidth resources in the networks calls for its efficient usage. To quantify this, bandwidth efficiency is defined as the ratio of the bandwidth utilized for data transmission to the total available bandwidth.
- *Quality of service support:* QoS in the networks defines the high mobility of the nodes. QoS provides the characteristics of ad-hoc networks.
- *Synchronization:* the sleep or wake up mode need to provide synchronization among the nodes. So, the synchronization is important for regulating the bandwidth reservation.
- *Hidden and exposed terminal problems:* The reason of two problems is the broadcast nature of the radio channel, namely, all the nodes within a node's transmission range receive its transmission.

4.4.1 Hidden Terminal Problem

Hidden terminal problem – as shown in Fig. 4.7 two nodes that are outside each-other's range perform simultaneous transmission to a node that is within the range of each of them, hence, there is a packet collision.

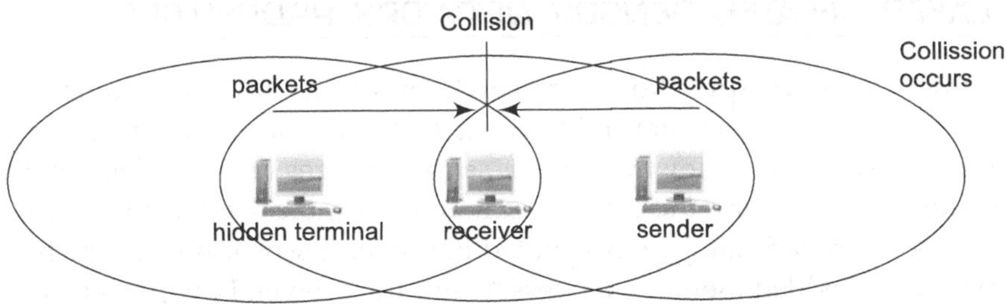

Fig. 4.7 Example of the hidden terminal problem

4.4.2 Exposed Terminal Problem

Exposed terminal problem – as shown in Fig. 4.8 the node is within the range of a node that is transmitting, and it cannot transmit to any node.

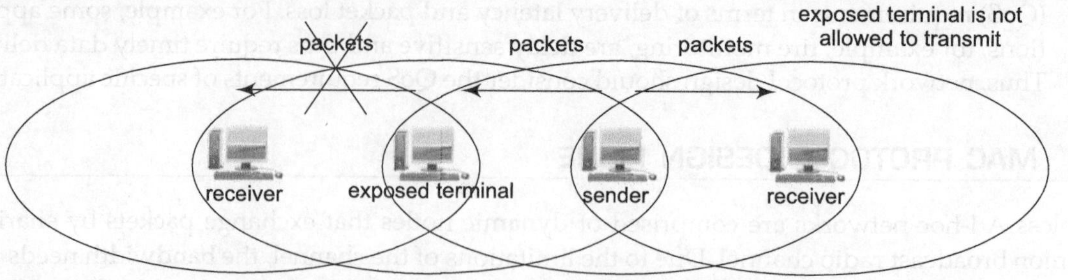

Fig. 4.8 Example of the exposed terminal problem

Hidden nodes mean increased probability of collision at a receiver, whereas exposed nodes may be denied channel access unnecessarily, which means underutilization of the bandwidth resources.

(*i*) *Error-prone shared broadcast channel:* In radio transmission, a node can listen to all traffic within its range and communicate without interference. When there is an access to the physical medium without session, there is high probability of collisions. The aim of a MAC protocol will be to minimize them, while maintaining fairness.

(*ii*) *No central coordination:* in ad-hoc networks, there is no central point of coordination due to the mobility of the nodes. Therefore, the control of the access to the channel must be distributed among them in order to exchange information between nodes. It is the responsibility of the MAC protocol to make sure this overhead is not a burden for the scarce bandwidth.

(*iii*) *Mobility of nodes:* The mobility of the nodes is one of its key features. The QoS reservations or the exchanged information might become useless, due to node mobility. The MAC protocol must be such that mobility has as little influence as possible on the performance of the whole network.

(*iv*) *Signal propagation delay:* Signal propagation delay is the amount of time needed for the transmission to reach the receiver. If the value of this parameter that is considered, while node may start transmitting, but it has not reached the node yet.

(*v*) *Hardware constraints:* Most radio-receivers are designed in such a way that only half duplex communication can take place. When a node is transmitting, the power level of the outgoing signal is higher than any received signal; therefore, the node receives its own transmission.

4.5 MAC LAYER RELATED SENSOR NETWORK PROPERTIES

Wireless sensor networks are appealing to researchers due to their wide range of application potential in areas such as target detection and tracking, environmental monitoring, industrial process monitoring, and tactical systems. However, lower sensing ranges result in dense networks, which bring the necessity to achieve an efficient medium access protocol subject to power constraints. Improvement in hardware technology has resulted in low-cost sensor nodes which are composed of a single chip with embedded memory, processor, and transceiver. Low power capacities lead to limited coverage and communication range for sensor nodes compared to other mobile devices.

However, the medium access decision within a dense network composed of nodes with low duty-cycles is a hard problem that must be solved in an energy-efficient manner. Under these circumstances, the MAC protocol must be energy efficient by reducing the potential energy wastes that are observed in sensor network applications. Maximizing the network lifetime is a research objective of sensor network, since sensor nodes assumed to be disposed when they are out of range of battery

4.5.1 Attributes of a Good MAC Protocol

To design a high-quality MAC protocol for the wireless sensor networks, the following attributes are to be considered. The first attribute is the energy efficient protocols in order to prolong the network lifetime. Other important attributes are scalability and adaptability to changes. Changes in network size, node density and topology should be handled rapidly and effectively for a successful adaptation. Some of the reasons behind these network property changes are limited node lifetime, addition of new nodes to the network and varying interference which may alter the connectivity and hence the network topology.

A good MAC protocol should gracefully accommodate such network changes. Other typical important attributes such as latency, throughput and bandwidth utilization may be secondary in sensor networks. Contrary to other wireless networks, fairness among sensor nodes is not usually a design goal, since all sensor nodes share a common task.

(*i*) Energy Efficiency: The sensor nodes are battery powered and it is often very difficult to change or recharge batteries for these sensor nodes. Sometimes it is beneficial to replace the sensor node rather than recharging them.

(*ii*) Latency: Latency requirement basically depends on the application. In the sensor network applications, the information exchange reported to the sink node in real time scenarios need to control and minimize.

(*iii*) Throughput: Throughput requirement also varies with different applications. Some of the sensor network application requires sampling the information with fine temporal resolution. In such sensor applications it is better that sink node receives more data.

(*iv*) Fairness: In many sensor network applications when bandwidth is limited, it is necessary to ensure that the sink node receives information from all sensor nodes fairly. However, among all of the above aspects the energy efficiency and throughput are the major aspects. Energy efficiency can be increased by minimizing the energy wastage.

4.5.2 MAC Performance Matrices

In order to evaluate and compare the performance of energy conscious MAC protocols, the following matrices are being used by the research community.

(*i*) Energy Consumption per bit: The energy efficiency of the sensor nodes can be defined as the total energy consumed/total bits transmitted. The unit of energy efficiency is joules/bit. The lesser the number, the better is the efficiency of a protocol in transmitting the information in the network. This performance matrices gets affected by all the major sources of energy waste in wireless sensor network such as idle listening, collisions, control packet overhead and overhearing.

(*ii*) Average Delivery Ratio: The average packet delivery ratio is the number of packets received to the number of packets sent averaged over all the nodes.

(*iii*) Average Packet Latency: The average packet latency is the average time taken by the packets to reach to the sink node.

(*iv*) Network Throughput: The network throughput is defined as the total number of packets delivered at the sink node per time unit.

4.5.3 Design Goals for Efficient MAC Protocols

The main design goals for the MAC protocols of sensor networks are to trade-off the following characteristics.

(*i*) Scalability: It is envisioned that most applications of sensor networks will involve large number of nodes. Therefore, scalability of the employed protocols is crucial. The resources, *i.e.*, time and bandwidth, sharing method and the arbitration strategy have to allow for fair access to the medium and to prevent excessive collisions. It is worth noting that the scalability of the link layer protocols is influenced by the network architecture and routing methodology.

(*ii*) Delay-predictability: A number of applications of sensor networks such as target tracking require delay-bounded delivery of data. Ensuing timeliness of data reception is typically handled at multiple layers in the communication stack.

(*iii*) Adaptability: In most applications of sensor networks traffic density varies significantly over time and from part of the network to another. So, the system should adjust to the changing mode to without failing the links.

(*iv*) Energy-efficiency: Energy is a scarce resource for sensor networks. The output power of the radio transmitter is directly proportional to distance squared and can significantly magnify in a noisy environment. Energy-conscious medium access control (MAC) can save transmission and reception energy by limiting the potential for collisions, minimizing the use of control messages, utilizing most of the available frequency band to shorten the transmission time, turning the radio into low power sleep mode when it is idle and finally, avoiding the excessive transitions among active and sleep states.

(*v*) Reliability: Reliable delivery of data is a classical design goal for all network infrastructures. Guaranteed packet delivery is ensured by the careful selection of error free links, avoidance of overloaded nodes, and the detection and the recovery from packet drops. There is usually a trade-off between the control traffic overhead and the level of reliability.

4.5.4 Design Principles

Design principles for a MAC protocol in ad-hoc network design principles that need to adopt and followed for effective implementation of protocol.

(*i*) The operation of the protocol should be distributed through all the nodes. The protocol should provide QoS support for real-time traffic.

(*ii*) The average delay for packet transmission should be as small as possible. The bandwidth should be utilized efficiently.

(*iii*) Each node must have a fair share of the available bandwidth. Control overhead should be minimized.

(*iv*) The hidden and exposed terminal problems should be minimized. The protocol must be scalable to large networks.

(*v*) Power control mechanisms are needed for efficient management of the energy consumption of the nodes.

(*vi*) Adaptive data rate controls the rate of outgoing traffic in relation to the network load and to the status of the other nodes.

(*vii*) Directional antennas are encouraged, the advantages are reduced interference, increased spectrum reuse, and reduced power consumption.

(*viii*) Time synchronization between the nodes should be provided.

4.5.5 Classification of MAC protocols for ad-hoc networks

As shown in Fig. 4.9 several criteria can be used for the classification of MAC protocols, such as time synchronization, initiation approach, and reservation approach. Ad-hoc network protocols can be classified into three basic types:

(*i*) Contention-based protocols;
(*ii*) Contention-based protocols with reservation mechanisms;
(*iii*) Contention-based protocols with scheduling mechanisms;

There are also some MAC protocols outside the above categories.

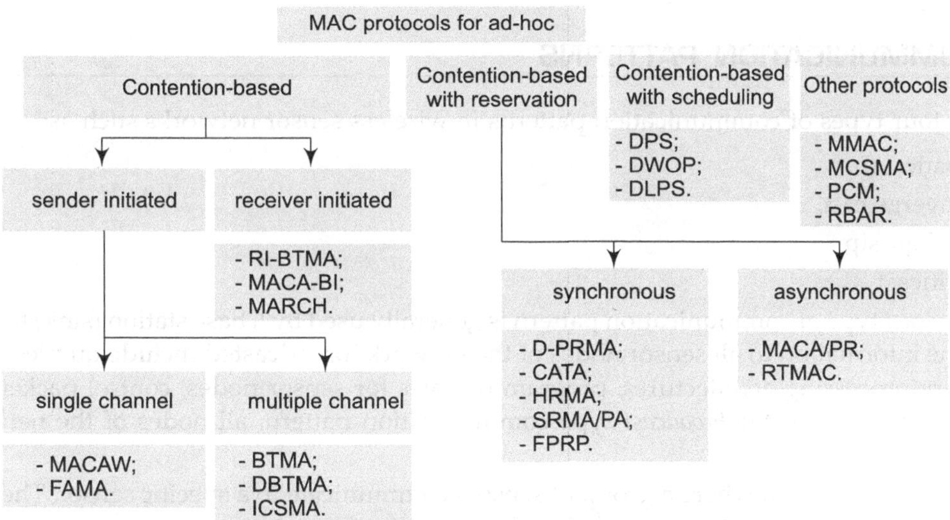

Fig. 4.9 Classification of the MAC protocols for ad-hoc networks

4.5.5.1 General definition of contention-based protocols

Here, the channel access policy is based on competition. Whenever a node needs to send a packet, it tries to get access to the channel. These protocols cannot provide QoS, since access to the network cannot be guaranteed beforehand. Contention-based MAC protocols are mainly based on the Carrier Sense Multiple Access (CSMA) or Carrier Sense Multiple Access/ Collision Avoidance (CSMA/CA).

The core idea is when one node needs to send data it will compete for wireless channel.

Contention-based protocols require no coordination among the nodes accessing the channel. Colliding nodes will back off for a random duration of time before attempting to access the channel.

The typical contention-based MAC protocols are S-MAC, T-MAC and UMAC.

4.5.5.2 General Definition of Contention-Based Protocols with Reservation Mechanisms

These protocols provide bandwidth reservation ahead; therefore, they can provide QoS support.

These can be further subdivided into:

- Synchronous protocols: there is time synchronization among all nodes in the network; the nodes in the neighborhood are informed of the reservations;
- Asynchronous protocols: no global synchronization is needed. Relative time is used for the reservations.

4.5.5.3 General Definition of Contention-Based Protocols with Scheduling Mechanisms

There can be packet scheduling at the nodes, or node scheduling for access to the channel. Node scheduling should not treat the nodes unfairly. Some of these protocols consider battery power in their node scheduling.

4.6 COMMUNICATION PATTERNS

There are four types of communication patterns in wireless sensor networks such as:

- Broadcast,
- Converge cast,
- Local gossip,
- Multicast

(*i*) *Broadcast* type of communication pattern is generally used by a base station (sink) to transmit some information to all sensor nodes of the network. Broadcasted include queries of sensor query-processing architectures, program updates for sensor nodes, control packets for the whole system. *In the broadcast* type communication pattern, all nodes of the network are intended receivers.

(*ii*) C*onverge cast* is that where a group of sensors communicate to a specific sensor. The destination node could be a cluster head, data fusion center, base station.

(*iii*) *Local gossip* is the detection of intruder by sensors. In some scenarios, the sensors that detect an intruder communicate with each other locally. This kind of communication pattern is called *local gossip*, where a sensor sends a message to its neighboring nodes within a range. The sensors that detect the intruder, then, need to send what they perceive to the information center.

(*iv*) *Multicast*, where a sensor sends a message to a specific subset of sensors. In this protocols clustering, cluster heads communicate with their members and thus the intended receivers may not be all neighbors of the cluster head, but just a subset of the neighbors.

4.7 CAUSE OF ENERGY WASTE

When a receiver node receives more than one packet at the same time, these packets are collide and they called "collided packets". These packets that cause the *collision have* to be discarded and the retransmissions of these packets are required which increase the energy consumption. Under these circumstances, the MAC protocol must be energy efficient by reducing the potential energy wastes that are observed in sensor network applications. Major sources of energy waste in wireless sensor network are basically of four types.

(*i*) Collision: Collision occurs when two nodes transmit at the same time and interfere with each other's transmission. Hence, re-transmissions increase energy consumption. If a transmitted packet is corrupted due to interference, it has to be discarded. The retransmissions increase energy consumption. Also, collision increases latency.

(*ii*) Overhearing: The overhearing, meaning that node picks up packets that are destined to other nodes. It means that there is no meaningful activity when nodes receive packets or a part of packets that are destined to other nodes.

(*iii*) Control Packet Overhead: The third source is control packet overhead. Sending and receiving control packets consumes energy too and less useful data packets can be transmitted. Control packet overhead such as RTS/CTS/ACK can be significant for wireless sensor networks that use small data packets.

(*iv*) Idle listening: Idle listening is the cost of actively listening for potential packets. Because nodes must keep their radio in receive mode, this source causes inefficient use of energy. If nothing is sensed, the sensor node will be in idle state for most of the time.

(*v*) Over emitting: The last reason for energy waste is over emitting, which is caused by the transmission of a message when the destination node is not ready.

The main goal of any MAC protocol for sensor network is to prevent the wastes. It is to minimize the energy waste due to idle listening, overhearing and collision. This is especially true in many sensor network applications.

4.8 PROPOSED MAC PROTOCOLS FOR WSN

Many medium access control (MAC) protocols for wireless sensor networks have been proposed in the recent years. Most of these protocols have energy conservation as an objective. The medium access control protocols for the wireless sensor networks can be classified broadly into two categories:

1. Schedule based and
2. Contention based.

The schedule based protocol can avoid collisions, overhearing and idle listening by scheduling transmit & listen periods but have strict time synchronization requirements.

The contention based protocols on the other hand relax time synchronization requirements and can easily adjust to the topology changes as some new nodes may join and others may die few years after deployment.

The protocols that are based on Carrier Sense Multiple Access (CSMA) technique have higher costs for message collisions, overhearing and idle listening. There is wide range of MAC protocols that define the essential behavior of the wireless sensor protocols. MAC protocols can be classified from four perspectives such as contention-based, TDMA-based, hybrid, and cross layer MAC. The following are the wide range of MAC protocols which are defined for sensor networks.

(*i*) Sensor-MAC (S-MAC)

(*ii*) Wise MAC

(*iii*) SIFT

(*iv*) Timeout-MAC (T-MAC) / Dynamic Sensor-MAC (DSMAC)

(*v*) Traffic-Adaptive MAC Protocol (TRAMA)
(*vi*) Berkeley a Access Control (B-MAC)
(*vii*) PAMAS: Power Aware Multi-Access Signaling
(*viii*) Optimized MAC
(*ix*) Data Gathering MAC (D-MAC)
(*x*) Self Organizing Medium Access Control for Sensor Networks (SMACS)
(*xi*) Energy Aware TDMA Based MAC
(*xii*) IEEE 802.11
(*xiii*) Aloha with Preamble Sampling

Some of the main MAC protocol of wireless sensor network is mentioned below.

4.8.1 Sensor-MAC (S-MAC)

S-MAC is a medium-access control (MAC) protocol designed for wireless sensor networks is a low power RTS-CTS based MAC protocol that makes use of loose synchronization between nodes to allow for duty cycling in sensor networks. The protocol uses three techniques to achieve low power duty cycling: periodic sleep, virtual clustering, and adaptive listening. It reduces energy consumption by allowing the nodes to periodically turn off their radio receivers and enter a low power sleep state. The duty cycle of a node is the ratio of the time it is awake to the total time. The lower the duty cycle, the lower is the power consumption of a sensor node.

Figure 4.10 shows the basic S-MAC scheme, where node 1 transmits data to node 2. A lot of energy is still wasted in this protocol during listen period as the sensor will be awake even if there is no reception/transmission.

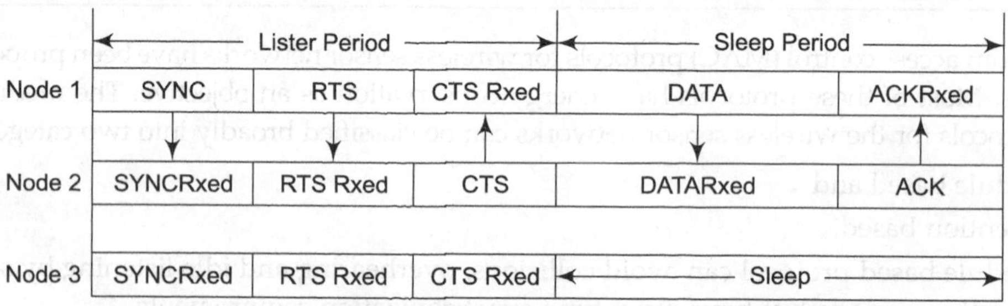

Fig. 4.10 Basic S-MAC scheme, node 1 transmits data to node 2

In S-MAC the channel access is contention based, using a scheme similar to the IEEE802.11 distributed coordination function. Locally managed synchronizations and periodic sleep listen schedules is based on these synchronizations where, node overhears its neighbor's transmissions wakes up for a short time at the end of the transmission. And if the node is the next-hop node, then its neighbor could pass data immediately. The end of the transmissions is known by the *duration* field of RTS/CTS packets. Neighboring nodes form virtual clusters to set up a common sleep schedule. If two neighboring nodes reside in two different virtual clusters, they wake up at listen periods of both clusters.

Advantages: The energy waste caused by idle listening is reduced by sleep schedules. In addition to its implementation simplicity, time synchronization overhead may be prevented with sleep schedule announcements.

Disadvantages: Broadcast data packets do not use RTS/CTS which increases collision probability. Adaptive listening incurs overhearing or idle listening if the packet is not destined to the listening node. Sleep and listen periods are predefined and constant, which decreases the efficiency of the algorithm under variable traffic load.

4.8.2 Wise MAC

WiseMAC protocol proposed by Enz et al. uses non-persistent CSMA (np-CSMA) with preamble sampling to decrease idle listening. WiseMAC is a CSMA based protocol using the preamble sampling technique to reduce the cost of idle listening. In the preamble sampling technique, a preamble precedes each data packet for alerting the receiving node. All nodes in a network sample the medium with a common period, but their relative schedule offsets are independent. If a node finds the medium busy after it wakes up and samples the medium, it continues to listen until it receives a data packet or the medium becomes idle again. The size of the preamble is initially set to be equal to the sampling period.

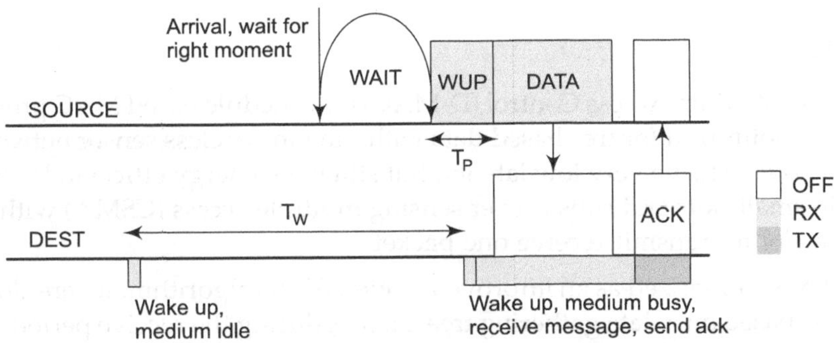

Fig. 4.11 Wise MAC protocol

The wake-up preamble introduces power consumption overhead both in transmission and in reception. To minimize this overhead, sensor nodes learn the offset between the sampling schedule of their direct neighbors and their own one. Knowing the sampling schedule of the destination, sensor nodes send messages just at the right time with a wake-up preamble of minimized length *TP* in the above Fig. 4.11. WiseMAC offers a method to dynamically determine the length of the preamble. That method uses the knowledge of the sleep schedules of the transmitter node's direct neighbors. Based on neighbors' sleep schedule table, WiseMAC schedules transmissions so that the destination node's sampling time corresponds to the middle of the sender's preamble. To decrease the possibility of collisions caused by that specific start time of wake-up preamble, a random wake-up preamble is advised.

Advantages: The simulation results show that WiseMAC performs better than one of the S-MAC variants. Besides, its dynamic preamble length adjustment results in better performance under variable traffic conditions.

Disadvantages: Main drawback of WiseMAC is that decentralized sleep-listen scheduling results in different sleep and wake-up times for each neighbor of a node. In addition, the hidden terminal problem exists. This is because WiseMAC is also based on non-persistent CSMA.

4.8.3 Traffic-Adaptive MAC Protocol (TRAMA)

TRAMA is a TDMA-based algorithm and proposed to increase the utilization of classical TDMA in an energy efficient manner. Time is divided into random-access and scheduled-access (transmission) periods. Random-access period is used to establish two-hop topology information where channel access is contention-based. The node announces the slots it will use as well as the intended receivers for these slots with a schedule packet. Bits correspond to one-hop neighbors ordered by their identities. Since, the receivers of those messages have the exact list and identities of the one- hop neighbors, they find out the intended receiver. When the vacant slots are announced, potential senders are evaluated for re-use of those slots. Priority of a node on a slot is calculated with a hash function of node's and slot's identities.

Advantages: In this protocol higher percentage of sleep time and less collision probability is achieved compared to CSMA based protocols. Since, intended receivers are having less communication to perform for multicast and broadcast type of communication patterns compared other protocols.

Disadvantages: Transmission slots are set to be seven times longer than the random access period. All the nodes are defined to be either in receive or transmit states so the duty cycle is at least considerably high.

4.8.4 D-MAC

The Data-Gathering Medium Access Control (D-MAC) is a schedule based MAC protocol which has been designed and optimized for tree based data gathering in wireless sensor network. The principal aim of DMAC is to achieve very low latency, but still to be energy efficient. In this protocol the time is divided in small slots and runs carrier sensing multiple access (CSMA) with acknowledgement within each slot to transmit/receive one packet.

D-MAC could be summarized as an improved Slotted Aloha algorithm, where slots are assigned to the sets of nodes based on a data gathering tree. Hence, during the receive period of a node, all of its child nodes has transmit periods and contend for the medium.

The sensor node periodically executes the basic sequence of '1' transmit, '1' receive and '*n*' sleep slots. In this approach a single packet from a source node at depth '*k*' in the tree reaches the sink node with a delay of just '*k*' time slots. This delay is very small and it is in the order of tens of milliseconds. A data gathering tree with staggered.

D-MAC includes an overflow mechanism to handle the problem when each single source node has low traffic rate, but the aggregate rate at intermediate node is larger than the basic duty cycle. In this mechanism, the sensor node will remain awake for one extra time slot after forwarding the packet is shown in Fig. 4.12.

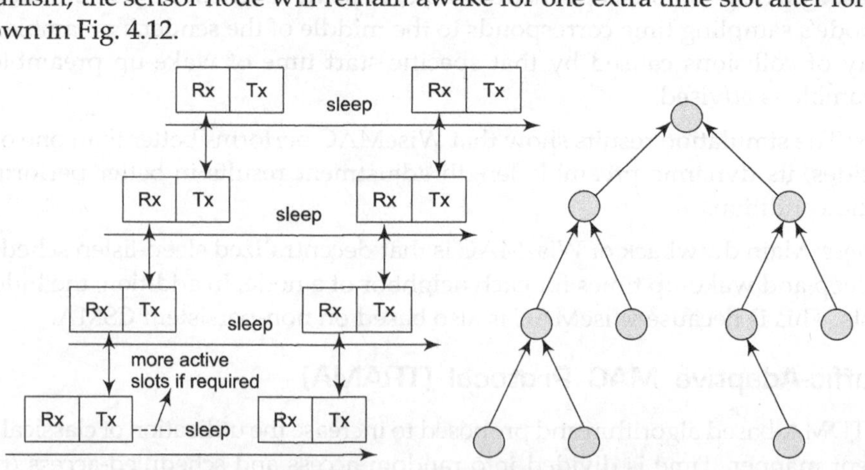

Fig. 4.12 Data gathering tree in D-MAC scheme

Advantages: D-MAC achieves low latency compared to other sleep/listen period assignment methods. The latency of the network is crucial for certain scenarios, in which D-MAC could be a strong candidate.

Disadvantages: Collision avoidance methods are not utilized, hence when a number of nodes that has the same schedule try to send to the same node, collisions will occur. This is a possible scenario in event-triggered sensor networks. Besides, the data transmission paths may not be known in advance, which precludes the formation of the data gathering tree.

4.8.5 SIFT

Sift is a MAC protocol proposed by Jamieson et al. is for event-driven sensor network environments. The motivation behind Sift is that when an event is sensed, and relayed with low latency. If no node starts to transmit in the first slot of the window, then each node increases its transmission probability exponentially for the next slot assuming that the number of competing nodes is small.

Sift is compared with 802.11 MAC protocol and it is show that Sift decreases latency considerably when there are many nodes trying to send a report. Since, Sift is a method for contention slot assignment algorithm, it is proposed to co-exist with other MAC protocols like S-MAC.

Advantages: Very low latency is achieved with many traffic sources. Energy consumption is traded off for latency as indicated below. It tuned to incur less energy consumption.

Disadvantages: One of the main drawbacks is increased idle listening caused by listening to all slots before sending. The second drawback is increased overhearing.

4.8.6 Timeout-MAC (T-MAC)

Timeout T-MAC is the protocol based on the S-MAC protocol in which the Active period is preempted and the sensor goes to the sleep period if no activation event has occurred for a time 'Ta' as shown in Fig. 4.13. The event can be reception of data, start of listen/sleep frame time etc. The time 'Ta' is the minimal amount of idle listening per frame.

Static sleep-listen periods of S-MAC result in high latency and lower throughput as indicated earlier. Timeout- MAC (T-MAC) is proposed to enhance the poor results of S-MAC protocol under variable traffic load. In T-MAC, listen period ends when no activation event has occurred for a time threshold TA.

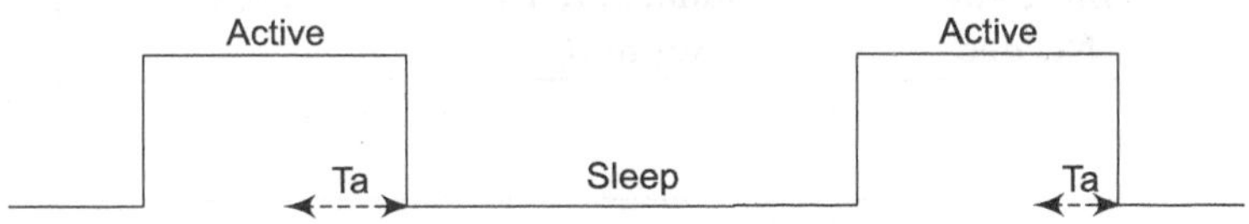

Fig. 4.13 Basic T-MAC Scheme

The interval Ta > Tci + Trt + Tta + Tct, where Tci is the length of the contention interval, Trt is the length of an RTS packet, Tta is the turn-around time (time between the end of the RTS packet and the beginning of the CTS packet) and Tct is the length of the CTS packet.

The energy consumption in the Timeout T- MAC protocol is less than the Sensor S-MAC protocol. But, the Timeout T-MAC protocol has high latency as compared to the S-MAC protocol.

The decision for TA is presented along with some solutions to the early sleeping problem defined in. T-MAC gives better results under the variable loads. One of the disadvantages is sleeping problem, because the synchronization of the listen periods within virtual clusters is broken.

4.8.7 Dynamic Sensor-MAC (DSMAC)

Dynamic Sensor-MAC (DSMAC) adds dynamic duty cycle feature to S-MAC. The aim is to decrease the latency for delay-sensitive applications. Within the SYNC period, all nodes share their one-hop latency values (time between the reception of a packet into the queue and its transmission). All nodes start with the same duty cycle. When a receiver node notices that average one-hop latency value is high, it decides to shorten its sleep time and announces it within SYNC period. Accordingly, after a sender node receives this sleep period decrement signal, it checks its queue for packets destined to that receiver node.

The latency observed with DSMAC is better than the one observed with S-MAC. Moreover, it is also shown to have better average power consumption per packet.

4.8.8 IEEE 802.11

The IEEE 802.11 is a well known contention based medium access control protocol which uses carrier sensing and randomized back-offs to avoid collisions of the data packets. The Power Save Mode (PSM) of the IEEE 802.11 protocol reduces the idle listening by periodically entering into the sleep state. Because of the problems in clock synchronization, neighbor discovery and network partitioning. This PSM mode is for the single-hop network, where the time synchronization is simple and may not be suitable for multi-hop networks.

4.8.9 Hybrid Contention-based and TDMA-based MAC

In recent years, there have been some hybrid proposals, which combine the advantages of contention-based MAC with that of TDMA-based MAC. All these protocols divide the access channel into two parts. Control packets are sent in the random access channel, and data packets are transmitted in the scheduled channel. The control channel schedules the data access. The hybrid protocols can gain high energy savings and offer better scalability and flexibility than any of contention-based MAC or TDMA-based MAC. Recently main hybrid protocols include Z-MAC, A-MAC, and IEEE 802.15.4. The Table 4.2 represents a comparison of MAC protocols investigated. The two S-MAC variants, namely, T-MAC and DSMAC, have the same features with S-MAC.

Table 4.2 Comparison of MAC protocols

	Time Synch. Needed	Comm. Pattern Support	Type	Adaptively to Changes
S-MAC/ T-MAC/ DSMAC	No	All	CSMA	Good
WiseMAC	No	All	np-CDMA	Good
TRAMA	Yes	All	TDMA/CSMA	Good
SIFT	No	All	CSMA/CA	Good
DMAC	Yes	Convergecast	TDMA/Slotted Aloha	Weak

Recently several medium access control protocols for the wireless sensor network have been proposed by the researchers. However, no protocol is accepted as standard. This is because the MAC protocol in general will be application specific. Therefore, there will not be one standard MAC protocol for the WSNs.

(*i*) The schedule based (TDMA) have collision free access to the medium, but the synchronization is critical. Moreover, there is difficulty in adapting to the changes in the network topology because of the addition and deletion of nodes.

(*ii*) The contention based (CSMA) have low latency and high throughput. However, it still suffers from the collisions.

(*iii*) The Frequency Division Multiple Access (FDMA) scheme also allow collision free access to the media, but the extra circuitry required to dynamically communicate with different radio channels increases the cost of the sensor nodes. This contradicts the main objective of the wireless sensor networks (WSNs).

(*iv*) The Code Division Multiple Access (CDMA) scheme also offers collision free access to the medium. However, the high computational complexity is the limitation in the lower energy consumption needs of the sensor network.

4.9 THE RELATED MAC PROTOCOLS IN WSN

While the traditional MAC protocols are designed to maximize the network throughput minimize the transmission latency and provide fairness transmission, the protocol design for wireless sensor networks focuses on minimizing the energy consumption among sensor nodes.

Time division multiple access (TDMA) and carrier sense multiple access (CSMA) are the familiar foundations for MAC protocols in wireless networks. TDMA-based protocols are naturally energy preserving, because they have time slots built-in, and do not suffer from collisions. TDMA divides time into small time slots and a node only wakes up at the indicated time slots in a cycle and goes into sleeping mode to save energy at the remaining time slots of the cycle. Therefore, TDMA provides contention free time slots to improve channel utilization. However, to maintain a TDMA schedule in the network is not an easy task. It requires more complexity in the nodes and needs an accurate time synchronization protocol which leads extra overheads. Furthermore, to allocate TDMA time slots is a complex problem that requires coordination.

CSMA-based protocols are the contention-based methods. Nodes contend the channel with its neighbors. Although, the random back off process is used, the collision might be even more frequent when the node density is high. The advantages of contention-based protocols are normally simple to implement, and scalable and adaptable to the heterogeneous deployments. In order to conserve energy, CSMA-based protocols must combine with duty cycle scheduling structure which allows nodes tune into sleeping mode for energy saving.

Most of the hybrid-based protocols combine the features of the TDMA-based and CSMA-based schemes and making themselves more efficient and adaptable to the variable environments in the wireless sensor networks. Although, a hybrid-based protocol can keep the advantages of both TDMA and CSMA schemes, it simultaneously has the drawbacks of the TDMA-based and CSMA-based schemes such as the synchronization overhead and the effects of the contention and the collision. In addition, a hybrid-based scheme is more complex and hard to implement than the individual scheme.

4.10 CROSS LAYER MAC FOCUSED ON ENERGY EFFICIENCY

Energy efficient MAC protocols presented above all focused on the design on the single MAC layer, without considering the correlation of other layers in WSN. Traditional network protocol stack is simple, however, it result in poor flexibility and low efficiency. With respect to the single layer protocols, cross layer is a novel method to improve energy efficiency.

In recent years cross layer MAC researches have been conducted for energy efficiency. This design takes into account, in a joint manner, the characteristics of the physical and the MAC layers. In a cross layer design, the use of forward error correction (FEC) coding and the determination of the awake/sleep periods for narrowband wireless sensor networks is presented. In a cross-layer mechanism, the protocol reduce control traffic routing overhead. In cross layer design, the useful information should be shared and transmitted among protocol layer to support the optimization mechanisms. MAC layer is influence physical layer by changing its transmission power and modulation. Routing layer chooses proper wireless links to relay packets to the destination so the routing decision will change the contention level at the MAC layer. Congestion and rate control in the transport layer will change the traffic volume in each communication link while traffic types will have great effect on MAC layer. So, cross layer method is a feasible way to improve the performances of MAC protocol and it need more research in depth.

SUMMARY

This chapter sets on giving a brief outline of the MAC protocols for ad-hoc wireless networks, focusing on contention-based algorithms with reservation and scheduling. In wireless sensor networks, how to reduce power consumption is one of the key design issues because sensor nodes are usually powered with capacity-limited batteries. Although, energy conservation can be addressed at each layer of the network protocol stack, we are focused on the MAC layer is explained in this chapter. There are some MAC protocols that have been discussed for wireless sensor networks. Typical examples include S-MAC, T-MAC, and H-MAC. To maximize the battery lifetime, sensor networks MAC protocols implement the variation of active/sleep mechanism. The IEEE 802.11 Distributed Coordinated Function (DCF), which is based on Carrier Sense Multiple Access with Collision Avoidance (CSMA/CA) Mechanism, is explained. In sensor networks, each node has a direct influence on its neighboring nodes while accessing the channel.

QUESTIONS

1. Explain the MAC Layer protocol.
2. Distinguish the difference between centralized and distributed topology.
3. Write a note on Distributed Coordination Function (DCF) Protocol.
4. What are the objectives of network design?
5. Brief out the MAC protocol design issue.
6. Illustrate the classification of the MAC protocols for ad-hoc networks.
7. How many types of communication patterns are there in wireless sensor networks?
8. What are the different types of MAC protocols which are defined for sensor networks?
9. Compare any MAC protocols with its feature.

BIBLIOGRAPHY

- C. Siva Ram Murthy and B. S. Manoj: "Ad-Hoc Wireless Networks: Architectures and Protocols".
- V. Kanodia, C. Li, A. Sabharwal, B. Sadeghi, and E. Knightly: Ordered Packet Scheduling in Wireless Ad-Hoc Networks: Mechanisms and Performance Analysis.

- Alexander Tyrrell, Gunther Auer and Christian Bettstetter: Firefly Synchronization in Ad-Hoc Networks.
- S.S., Kulkarni, "TDMA services for Sensor Networks", Proceedings of 24th International Conference on Distributed Computing Systems Workshops, Pages: 604 - 609, 23-24 March 2004.
- W. Ye, J. Heidemann, D. Estrin, "Medium Access Control With Coordinated Adaptive Sleeping for Wireless Sensor Networks", IEEE/ACM Transactions on Networking, Volume: 12, Issue: 3, Pages: 493 - 506, June 2004.
- V. Rajendran, K. Obraczka, J.J. Garcia-Luna-Aceves, "Energy- Efficient, Collision-Free Medium Access Control for Wireless Sensor Networks", Proc. ACM SenSys 03, Pages: 181 - 192, Los Angeles, California, 5-7 November 2003.
- G. Lu, B. Krishnamachari, C.S. Raghavendra, "An adaptive energy- efficient and low-latency MAC for data gathering in wireless sensor networks", Proceedings of 18th International Parallel and Distributed Processing Symposium, Pages: 224, 26-30 April 2004.
- P. Lin, C. Qiao, and X. Wang, "Medium access control with a dynamic duty cycle for sensor networks", IEEE Wireless Communications and Networking Conference, Volume: 3, Pages: 1534 - 1539, 21-25 March 2004.
- An Energy-Efficient MAC Protocol in Wireless Sensor Networks: A Game Theoretic Approach, S. Mehta and K. S. Kwak, EURASIP Journal on Wireless Communications and Networking, Volume 2010.
- Ye Wei, Herdemann John, Estin Deborah. An Energy-Efficient MAC Protocol for Wireless Sensor Networks, Proc. of the INFOCOM 2002, San.
- Suh Changsu, Ko Young-Bae, Son Dong-Min. An Energy Efficient Cross-Layer MAC Protocol for Wireless Sensor Networks[C]//Proc. of the eighth Asia Pacific Web conference, Harbin, 2006:410-419
- Liang Song, Dimitrios Hatzinakos. A Cross-Layer Architecture of Wireless Sensor Networks for Target Tracking[M]//IEEE/ACM Transactions on Networking, 2007,15:145-158
- Wei Ye, J.Heidemann and D. Estrin: An Energy- Efficient MAC Protocol for Wireless Sensor Networks, IEEE INFOCOM, New York, Vol. 2, pp. 1567-1576 (June 2002).
- IEEE Standard 802.11. Wireless LAN Medium Access Control (MAC) and Physical Layer (PHY) Specifications (1999).
- G. Lu, B. Krishnamachari, C. Raghavendra: An Adaptive Energy Efficient and Low Latency MAC for Data Gathering in Wireless Sensor Networks, Proceedings of 18thth International Parallel and Distributed Processing Symposium (April 2004).

5 Routing Protocol in WSN

5.1 INTRODUCTION

Wireless sensor network (WSNs) are widely used in electronics. It is a set of hundreds or thousands of micro sensor nodes that have capabilities of sensing, establishing wireless communication between each other and doing computational and processing operations.

In sensor networks, sensor nodes can operate as nodes, cluster heads, base stations, etc. It means, a sensor node must be heterogeneous. Fault tolerance must always be taken into account. Number of active nodes in a network is variable.

A high number of active nodes can cause infinite queues. Mobility tolerance in target detection and tracking as well as adaptive routing techniques are needed. Quality of service requires limited delays. All these challenge place routing on the sharp edge. The following points summarize the requirements for routing in wireless sensor networks:

I. Low power consumption,

II. Maximum network lifetime,

III. Low data rates,

IV. In-network stability,

V. Mobility tolerance.

Advantages can be discovered in mobile networks, where the topology changes frequently and where nodes can have intermittent connectivity. Routing is the act of moving information across an inter-network from a source to a destination. It's also referred to as the process of choosing a path over which to send the packets. Routing is often contrasted with bridging. The routing algorithm is the part of the network layer software responsible for deciding which output line an incoming packet should be transmitted on, *i.e.,* what should be the next intermediate node for the packet.

Wireless sensor networks are now used in many applications including military, environmental, healthcare applications, home automation and traffic control. A sensor network is composed of a large number of tiny autonomous devices, called sensor nodes that has limited sensing and computational capabilities to communicate only in short distances. Routing protocol is a set of rules defining the way router machines find the way that packets containing information have to follow to reach the intended destination.

5.2 STRUCTURE OF SENSOR NETWORK

A Wireless Sensor Network (WSN) contains hundreds or thousands of these sensor nodes. These sensors have the ability to communicate either among each other or directly to an external base-station (BS). A greater number of sensors allows for sensing over larger geographical regions with greater accuracy. Fig. 5.1 shows the structural view of sensor network.

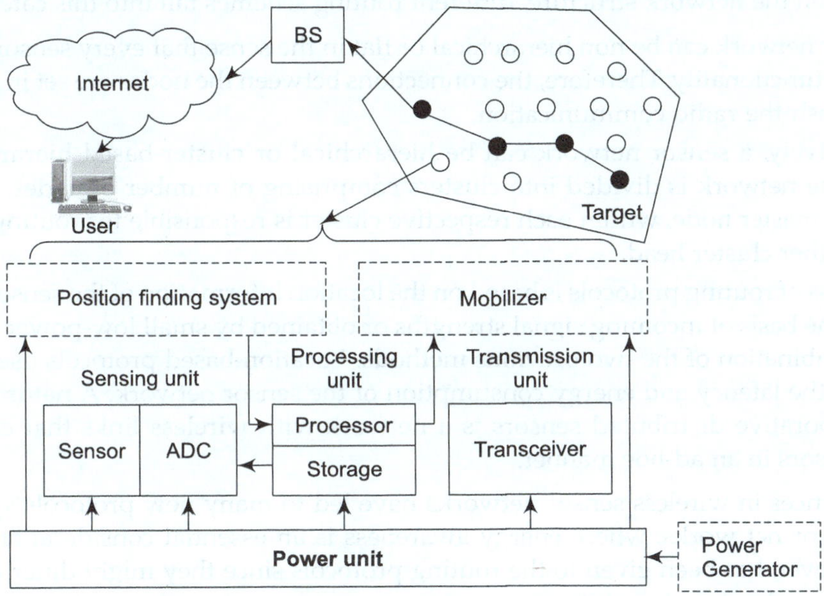

Fig. 5.1 Structural view of sensor network

Sensor nodes are usually scattered in a sensor field, which is an area, where the sensor nodes are deployed. Each scattered sensor nodes has the capability to collect and route data either to other sensors or back to an external base station. A base-station is a fixed node or a mobile node capable of connecting the sensor network to an existing communications infrastructure or to the Internet for information collection.

Routing in WSNs is very challenging due to the inherent characteristics that distinguish these networks from other wireless networks like mobile ad-hoc networks or cellular networks. First, due to the relatively large number of sensor nodes, the overhead of ID maintenance is high. Moreover, traditional IP-based protocols may not be applied to WSNs. In WSNs, getting the data is more important than knowing the IDs of which nodes sent the data. Second, in contrast to typical communication networks, almost all applications of sensor networks require the flow of sensed data from multiple sources to a particular BS. Third, sensor nodes are tightly constrained in terms of energy, processing, and storage capacities. Thus, they require careful resource management. Fourth, in most application scenarios, nodes in WSNs are generally static or dynamic nodes. So, sensor networks are application specific, where the design requirements of a sensor network change with application. Fifth position awareness of sensor nodes is data collection based on the location use Global Positioning System (GPS) hardware for this purpose.

Finally, data collected by many sensors in WSNs is typically based on common phenomena; hence there is a high probability that this data has some redundancy. Such redundancy needs to be exploited by the routing protocols to improve energy and bandwidth utilization.

5.3 CLASSIFICATION OF WSN ROUTING

In general, classification of a WSN routing methodology can be done into two main categories:-

I. Based on network structure or

II. Based on the protocol operation.

Depending on the network structure, different routing schemes fall into this category.

(*a*) A sensor network can be non hierarchical or flat in the sense that every sensor has the same role and functionality. Therefore, the connections between the nodes are set in short distance to establish the radio communication.

(*b*) Alternatively, a sensor network can be hierarchical or cluster-based hierarchical model, where the network is divided into clusters comprising of number of nodes. Cluster head, which is master node, within each respective cluster is responsible for routing the information to other cluster head.

Another class of routing protocols is based on the location information of the sensor nodes either estimated on the basis of incoming signal strengths or obtained by small low-power GPS receivers or even by combination of the two previous methods. Location-based protocols use this information to reduce the latency and energy consumption of the sensor network. A natural architecture for such collaborative distributed sensors is a network with wireless links that can be formed among the sensors in an ad-hoc manner.

Recent advances in wireless sensor networks have led to many new protocols specifically designed for sensor net works, where energy awareness is an essential consideration. Most of the attention, however, has been given to the routing protocols since they might diûer depending on the application and network architecture.

Sensor nodes are constrained in energy supply and bandwidth. Such constraints have posed many challenges of energy awareness to the design and management of large number sensor nodes in the sensor networks. Therefore, the research has been focused for energy-efficient route setup and reliable relaying of data from the sensor nodes to the sink so that the lifetime of the network is maximized.

Routing in sensor networks is very challenging due to several characteristics of communication and wireless ad-hoc networks deployment constraints.

I. First of all, it is not possible to build a global addressing scheme for the deployment of sheer number of sensor nodes. Therefore, classical IP-based protocols cannot be applied to sensor networks.

II. Second, in contrary to typical communication networks almost all applications of sensor networks require the flow of sensed data from multiple regions (sources) to a particular sink.

III. Third, generated data trafic has significant redundancy in it since multiple sensors may generate same data within the vicinity of a phenomenon. Such redundancy needs to be exploited by the routing protocols to improve energy and bandwidth utilization.

IV. Fourth, sensor nodes are tightly constrained in terms of transmission power, on-board energy, processing capacity and storage and thus require careful resource management.

Due to such difference, many new algorithms have been proposed for the problem of routing data in sensor networks. These routing mechanisms have considered the characteristics of sensor nodes along with the application and architecture requirements.

5.4 ROUTING DESIGN CHALLENGES

Wireless sensor networks consist of a huge number of sensor nodes. Mainly, all of them are battery powered. The main requirements for wireless sensor networks are low power consumption, long network lifetime, low data rates, in-network stability, mobility tolerance, scalability, etc. The routing protocol is a critical issue. Routing algorithms should always be power aware, because sensor network lifetime is equal to sensor node lifetime. The design of routing protocols in WSNs is influenced by certain challenging factors. These factors must be overcome for efficient communication in WSNs. The following, summarize some of the routing challenges and design issues that affect routing process in WSNs.

(*i*) Node deployment: Node deployment in WSNs can be either deterministic or randomized. In deterministic deployment, the sensors are manually placed and data is routed through predetermined paths. However, in random node deployment, the sensor nodes are scattered randomly creating an infrastructure in an ad-hoc manner. If the resultant distribution of nodes is not uniform, optimal clustering becomes necessary to allow connectivity and enable energy efficient network operation.

(*ii*) Energy consumption: sensor nodes use up their limited supply of energy performing computations and transmitting information in a wireless environment so such, sensor node is dependence on the battery lifetime. In a multihop WSN, sensor nodes prone to power failure due to significant topological changes.

(*iii*) Data collection Model: Data sensing and reporting in WSNs is dependent on the application and the time criticality of the data reporting. Data reporting is based on either time-driven (continuous), event-driven, query-driven, and hybrid. The time-driven model requires periodic data monitoring. In event-driven and query-driven models, sensor nodes collect data based on sudden and drastic changes. The routing protocol is highly influenced by the data reporting model with regard to energy consumption and route stability.

(*iv*) Node/Link Heterogeneity: In many studies, all sensor nodes were assumed to be homogeneous, *i.e.*, having equal capacity in terms of computation, communication, and power. The existence of heterogeneous set of sensors raises many technical issues related to data routing.

(*v*) Fault Tolerance: Sensor nodes normally may fail or blocked due to lack of power, physical damage, or environmental interference. The failure of sensor nodes should not affect the overall task of the sensor network. If many nodes fail, MAC and routing protocols must accommodate formation of new links and routes to the data collection base stations. Therefore, multiple levels of redundancy may be needed in a fault-tolerant sensor network.

(*vi*) Scalability: The number of sensor nodes deployed in the sensing area should be scalable enough to respond to events in the environment. In sensor network routing scheme must be able to work with this huge number of sensor nodes and should be scalable.

(*vii*) Network Dynamics: Most of the network architectures assume that sensor nodes can be either dynamic or static depending on the application, *e.g.*, it is dynamic in a target detection/ tracking application, while it is static in forest monitoring for early fire prevention. Monitoring static events allows the network to work in a reactive mode; where as dynamic events in most applications require periodic reporting.

(*viii*) Transmission Media: In a multi-hop wireless sensor network, communicating nodes are linked by a medium that is associated with wireless channel inferences such as fading, attenuation, high error rate etc for the operation of the sensor network. So, the protocol like TDMA, CSMA need to choose to control the bandwidth usage.

(*ix*) Connectivity: High node density in sensor networks is needed to be highly connective as per topology. The connectivity depends on the, possibly random, distribution of nodes from being shrinking due to sensor node failures.

(*x*) Coverage: In WSNs, each sensor node obtains a certain view of the environment with limited range and in accuracy. This physical area coverage is an important design parameter in WSNs.

(*xi*) Data Aggregation: Data aggregation is the combination of data from different sources according to a certain aggregation function, *e.g.*, duplicate suppression, minima, maxima and average. This technique has been used to achieve energy efficiency and data transfer optimization in a number of routing protocols.

(*xii*) Quality of Service: The data delivered to the destination node without error determines the QoS. Therefore, reduction in latency for data delivery for time-constrained applications is directly related to network lifetime. As the energy gets depleted, the network may be required to reduce the quality of the results in order to reduce the energy dissipation in the nodes and hence lengthen the total network lifetime.

(*xiii*) Reliability: Since, messages travel multiple hops it is important to have a high reliability on each link, otherwise the probability of a message transiting the entire network would be unacceptably low.

(*xiv*) Integration with wake/sleep schedules: To save power many WSN place nodes into sleep states. Obviously, an awake node should not choose an asleep node as the next hop.

(*xv*) Unicast, multicast and any cast semantics: As mentioned above, in most cases a WSN routes messages to a geographic destination. There are several possibilities.

- (*i*) First, the message may also include an ID with a specific unicast node in this area as the target, or the semantics may be that a single node closest to the geographic destination is to be the unicast node.
- (*ii*) Second, the multicast that could be that all nodes within some area around the destination address should receive the message.
- (*iii*) Third, for any node, called any cast, in the destination area to receive the message.

(*xvi*) Real-Time: For some applications, messages must arrive at a destination by a deadline. Due to the high degree of uncertainty in WSN it is difficult to develop routing algorithms with any guarantees. Velocity is a metric that combines the deadline and distance that a message must travel.

(*xvii*) Mobility: Routing is complicated if either the message source or destination or both are moving. Solutions include continuously updating local neighbor tables or identifying proxy nodes which are responsible for keeping track of where nodes are away from its original location.

(*xviii*) Voids: Since, WSN nodes have a limited transmission range, it is possible that for some nodes in the routing path there are no forwarding nodes in the direction a message is supposed to travel. Protocols like GPSR solve this problem by choosing some other node "not" in the correct direction in an effort to find a path around the void.

(*xix*) Security: If adversaries exist, they can perpetrate a wide variety of attacks on the routing algorithm including selective forwarding, black hole, Sybil, replays, wormhole and denial of service attacks. Unfortunately, almost all WSN routing algorithms have ignored security and are vulnerable to these attacks.

(*xx*) Congestion: Today, many WSN have periodic or infrequent traffic. However, congestion is a problem for more demanding WSN and is expected to be a more prominent issue with larger systems that might process audio, video and have multiple base stations (creating more cross traffic). Even in systems with a single base station, congestion near the base station is a serious problem, since traffic converges at the base station. Solutions use backpressure, reducing collisions as possible which only exacerbate the congestion problem.

5.5 ROUTING PROTOCOL CATEGORY

Routing protocols are specific routing algorithms with properties used in different routing topology classes, such as flat, hierarchical and location-based routing. The three main categories of routing protocols are:

I. Data-centric, protocols are query-based and depend on the naming of desired data, which helps in eliminating many redundant transmissions.

II. Hierarchical protocols aim at clustering the nodes so that cluster heads can do some aggregation and reduction of data in order to save energy.

III. Location-based protocols utilize the position information to relay the data to the desired regions rather than the whole network.

IV. Flow modeling and protocols is the last category includes routing approaches that are based on general network that strive for meeting some quality of service (QoS) awareness requirements along with the routing function.

5.5.1 Data Centric Routing Protocol

In many applications of sensor networks, it is not feasible to assign global identifiers to each node due to the absolute number of nodes deployed. Such lack of global identification along with random deployment of sensor nodes makes it hard to select a specific set of sensor nodes to be queried. Therefore, data is usually transmitted from every sensor node within the deployment region with significant redundancy. Since, this is very ineficient in terms of energy consumption, routing protocols that will be able to select a set of sensor nodes and utilize data aggregation during the relaying of data. This has led to data centric routing, which is different from traditional address-based routing, where routes are created between addressable nodes managed in the network layer of the communication stack. In data-centric routing, the sink sends queries to certain regions and waits for data from the sensors located in the selected regions.

Data-centric routing protocol is following types:

I. SPIN

II. Flooding and gossiping

III. Directed Diffusion.

IV. Energy-aware routing

V. Rumor routing

VI. Gradient-based routing

VII. Constrained anisotropic diffusion routing (CADR)
VIII. COUGAR
IX. ACQUIRE

I. **Sensor protocols for information via negotiation (SPIN):** SPIN is the first data-centric protocol, which considers data negotiation between nodes in order to eliminate redundant data and save energy. It is the early work to pursue a data-centric routing mechanism. The idea behind SPIN is to name the data using high-level descriptors or meta-data. Directed diffusion has been the latest developed protocol in data-centric routing.

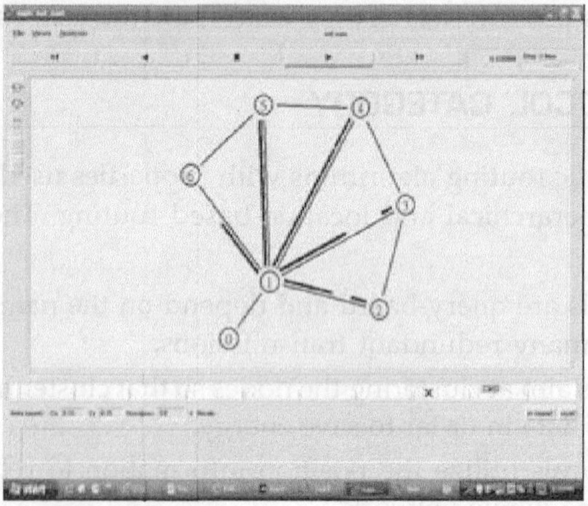

Fig. 5.2 Data-centric routing protocol simulation of SPIN-1.

The advantages of SPIN are that topological changes are localized, since each node needs to know only its single-hop neighbors.

The disadvantage of SPIN is that data advertisement mechanism cannot guarantee the delivery of data. For instance, if the nodes that are interested in the data are far away from the source node and the nodes between source and destination are not interested in that data, such data will not be delivered to the destination at all. Therefore, SPIN is not a good choice for applications such as intrusion detection, which require reliable delivery of data packets over regular intervals.

II. **Flooding and gossiping:** Flooding and gossiping are two classical mechanisms to relay data in sensor networks without the need for any routing algorithms and topology maintenance. In flooding, each sensor receiving a data packet broadcasts it to all of its neighbors and this process continues until the packet arrives at the destination or the maximum number of hops for the packet is reached. On the other hand, gossiping is a slightly enhanced version of flooding, where the receiving node sends the packet to a randomly selected neighbor, which picks another random neighbor to forward the packet to and so on.

III. **Directed Diffusion:** It is an important milestone in the data-centric routing research of sensor networks. The idea aims at diffusing data through sensor nodes by using a naming scheme for the data. The main reason behind using such a scheme is to get rid of unnecessary operations of network layer routing in order to save energy.

Advantages of data centric is, all communication is neighbor-to-neighbor with no need for a node addressing mechanism. Direct Diffusion is highly energy efficient, since it is on demand and there is no need for maintaining global network topology. However, Directed Diffusion cannot be applied to all sensor network applications, therefore is not a good choice as a routing protocol for the applications such as environmental monitoring.

IV. **Energy-aware routing:** This protocol was proposed by Shah and Rabaey to use a set of sub-optimal paths occasionally to increase the lifetime of the network. These paths are chosen by means of a probability function, which depends on the energy consumption of each path. The approach argues that using the minimum energy path all the time will deplete the energy of nodes on that path. Instead, one of the multiple paths is used with a certain probability so that the whole network lifetime increases as shown in Fig. 5.3.

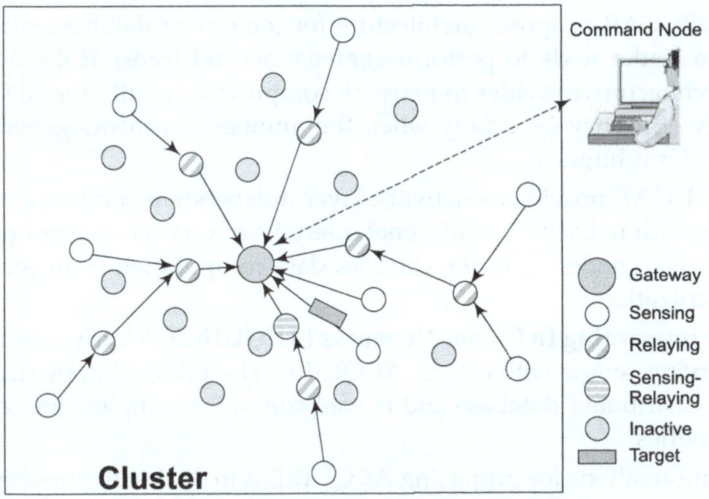

Fig. 5.3 A typical cluster in a sensor network

The protocol assumes that each node is addressable through a class-based addressing which includes the location and types of the nodes. There are 3 phases in the protocol:

(*a*) Setup phase
(*b*) Data communication phase
(*c*) Route maintenance phase

V. **Rumor routing:** This is another variation of Directed Diffusion and is mainly intended for contexts in which geographic routing criteria are not applicable. However, in some cases there is only a little amount of data requested from the nodes and thus the use of flooding is unnecessary. An alternative approach is to flood the events if number of events is small and number of queries is large. Rumor routing is between event flooding and query flooding. The idea is to route the queries to the nodes that have observed a particular event rather than flooding the entire network to retrieve information about the occurring events

However, rumor routing performs well only when the number of events is small. Rumor routing maintains only one path between source and destination as opposed to Directed Diffusion, where data can be sent through multiple paths at low rates.

VI. **Gradient-based routing (GBR):** Gradient-based routing protocol was proposed by Schurgers et al. is a slightly changed version of Directed Diûusion. The idea is to keep the number of hops when the interest is diffused through the network. Hence, each node can discover the minimum number of hops to the sink, which is called height of the node. The difference between a node's height and that of its neighbor is considered the gradient on that link. A packet is forwarded on a link with the largest gradient.

VII. **Constrained anisotropic diffusion routing (CADR):** In CADR, each node evaluates an information/cost objective and routes data based on the local information/cost gradient and end-user requirements. The information utility measure is modeled using standard estimation theory. CADR is a protocol, which strives to be a general form of Directed Diffusion. Two techniques namely information-driven sensor querying (IDSQ) and constrained anisotropic diffusion routing are proposed. The idea is to query sensors and route data in a network in order to maximize the information gain, while minimizing the latency and bandwidth. This is achieved by activating only the sensors that are close to a particular event and dynamically adjusting data routes.

VIII. **COUGAR:** COUGAR proposes architecture for the sensor database system, where sensor nodes select a leader node to perform aggregation and transmit the data to the gateway (sink). The architecture provides in network computation ability for all the sensor nodes to ensure energy-efficiency especially when the number of sensors generating and sending data to the leader is huge.

Although, COUGAR provides a network-layer independent solution for querying the sensors, it has some drawbacks: 1. additional query layer on each sensor node will bring extra overhead to sensor nodes, 2. In the network data computation from several nodes will require synchronization.

IX. **Active Query forwarding In Sensor Networks (ACQUIRE):** A fairly new data-centric mechanism for querying sensor networks is ACQUIRE. The protocol approach views the sensor network as a distributed database and is well-suited for complex queries which consist of several sub queries.

One of the main motivations for proposing ACQUIRE is to deal with one-shot, complex queries for data, where a response can be provided by many nodes.

5.5.2 Hierarchical Routing Protocol

It is the next category of routing protocols. In the hierarchical routing protocol, the base station is fixed and located far away from the sensors. Hierarchical routing is the procedure of arranging routers in a hierarchical manner. The basic idea of hierarchical routing protocol is to organize sensor nodes into cluster based on the received signal strength and use local cluster-heads as routers to base station. It performs local data fusion and aggregation at cluster-heads to further reduce energy consumption. Sensor nodes elect themselves to be local cluster heads with a certain probability. The non-cluster head node will join a cluster-head that requires minimum communication energy. Hierarchical routing protocols are following types:

I. Low-energy adaptive clustering hierarchy (LEACH)
II. Power-efficient Gathering in Sensor Information Systems (PEGASIS)
III. Threshold sensitive Energy Efficient sensor Network protocol (TEEN)
IV. Energy-aware routing for cluster-based sensor networks
V. Self-organizing protocol

I. **LEACH:** Low-energy adaptive clustering hierarchy (LEACH) is one of the most popular hierarchical routing algorithms for sensor networks. The idea is to form clusters of the sensor nodes based on the received signal strength and use local cluster heads as routers to the sink. This will save energy, since the transmissions will only be done by such cluster heads rather than all sensor nodes.

LEACH is completely distributed and requires no global knowledge of network. However, LEACH uses single-hop routing where each node can transmit directly to the cluster-head and the sink. Therefore, it is not applicable to networks deployed in large regions.

II. **PEGASIS and Hierarchical-PEGASIS**: Power-efficient Gathering in Sensor Information Systems (PEGASIS) is an improvement of the LEACH protocol. PEGASIS forms chains from sensor nodes so that each node transmits and receives from a neighbor and only one node is selected from that chain to transmit to the base station (sink). Gathered data moves from node to node, aggregated and eventually sent to the base station. Hierarchical-PEGASIS is an extension to PEGASIS, which aims at decreasing the delay incurred for packets during transmission to the base station In order to reduce the delay in PEGASIS, simultaneous transmissions of data messages are pursued.

III. **Threshold sensitive Energy Efficient sensor Network protocol (TEEN):** TEEN pursues a hierarchical approach along with the use of a data-centric mechanism. The sensor network architecture is based on a hierarchical grouping where closer nodes form clusters and this process goes until base station (sink) is reached. TEEN is a hierarchical protocol designed to be responsive to sudden changes in the sensed attributes such as temperature.

IV. **Energy-aware routing for cluster-based sensor networks:** This protocol was proposed by Younis et al. in a different hierarchical routing algorithm based on three-tier architecture. Sensors are grouped into clusters prior to network operation. The algorithm employs cluster heads, namely gateways, which are less energy constrained than sensors and assumed to know the location of sensor nodes. Gateways maintain the states of the sensors and sets up multi-hop routes for collecting sensors data. A TDMA based MAC is used for nodes to send data to the gateway. The gateway informs each node about slots in which it should listen to other nodes transmission and slots, which the node can use for its own transmission. The command node (sink) communicates only with the gateways.

V. **Self-organizing protocol:** This protocol was proposed by Subramanian and Katz. It describes a self-organizing protocol but develop taxonomy of sensor applications as well.

The architecture supports heterogeneous sensors that can be mobile or stationary. Some sensors, which can be either stationary or mobile, probe the environment and forward the data to designated set of nodes that act as routers. Router nodes are stationary and form the backbone for communication.

The disadvantage is in the organization phase of algorithm, which is not on-demand, therefore introducing extra overhead.

5.5.3 Location-based Routing Protocols

In most of the routing protocols for sensor networks it require location information for sensor nodes to calculate the distance between two particular nodes so that energy consumption can be estimated. Since, there is no addressing scheme for sensor networks like IP-addresses and they are spatially deployed on a region, location information can be utilized in routing data in an energy efficient way. For instance, if the region to be sensed is known, using the location of sensors, the query can be diffused only to that particular region which will eliminate the number of transmission significantly.

Location-based routing protocols are following types:

I. Minimum energy communication network (MECN)

II. Geographic and energy-aware routing (GEAR),
III. Geographic adaptive fidelity (GAF)

I. **MECN and SMECN:** Minimum energy communication network (MECN) sets up and maintains a minimum energy network for wireless networks by utilizing low power GPS. A minimum power topology for stationary nodes including a master node is found. MECN assumes a master-site as the information sink, which is always the case for sensor networks. MECN is self-reconûguring and thus can dynamically adapt to node s failure or the deployment of new sensors.

The small minimum energy communication network (SMECN) is an extension to MECN. In MECN, it is assumed that every node can transmit to every other node, which is not possible every time. In SMECN possible obstacles between any pair of nodes are considered.

II. **GEAR:** This protocol was proposed by Yu et al. have suggested the use of geographic information while disseminating queries to appropriate regions since data queries often includes geographic attributes. The protocol, namely geographic and energy-aware routing (GEAR), uses energy aware and geographically informed neighbor selection heuristics to route a packet towards the target region.

GEAR compliments Directed Delusion in this way and thus conserves more energy. In GEAR, each node keeps an estimated cost and a learning cost of reaching the destination through its neighbors.

III. **GAF: Geographic adaptive fidelity (GAF)** is an energy-aware location-based routing algorithm designed primarily for mobile ad-hoc networks, but may be applicable to sensor networks as well. GAF conserves energy by turning-oû unnecessary.

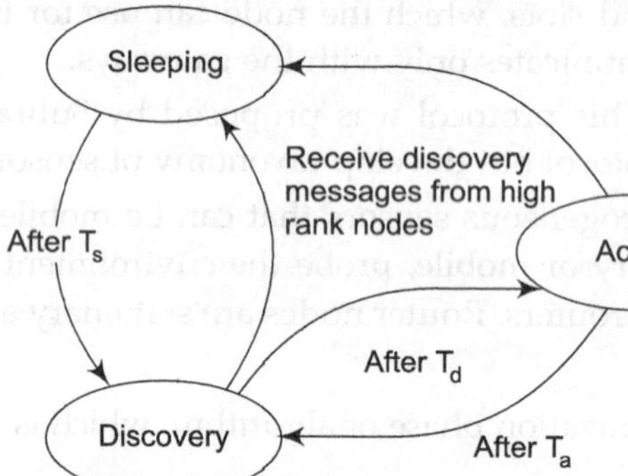

Fig. 5.4 State transitions in GAF

Each node uses its GPS-indicated location to associate itself with a point in the virtual grid. Nodes change states from sleeping to active in turn so that the load is balanced. There are three states deûned in GAF as shown in Fig. 5.4. These states are discovery, for determining the neighbors in the grid, active reflecting participation in routing and sleep when the radio is turned off. Thus, GAF can substantially increase the network lifetime as the number of nodes increases.

5.5.4 Network Flow and QoS-aware Protocols

The routing protocols proposed for sensor networks pursue somewhat different approach such as network flow and QoS. In some approaches, route setup is modeled and solved as a network flow problem. QoS-aware protocols consider end-to-end delay requirements while setting up the paths in the sensor network.

Network flow and QoS-aware protocols are following types:

I. Maximum lifetime energy routing
II. Maximum lifetime data gathering
III. Minimum cost forwarding
IV. Sequential assignment routing (SAR)
V. Energy-aware QoS routing protocol
VI. SPEED

I. **Maximum lifetime energy routing:** This protocol was proposed by Chang and Tassiulas presents an interesting solution to the problem of routing in sensor net-works based on a network how approach. The main objective of the approach is to maximize the network lifetime by carefully defining link cost as a function of node remaining energy and the required transmission energy using that link. Finding traffic distribution is a possible solution to the routing problem in sensor networks and based on that, comes the name "maximum lifetime energy routing". The solution to this problem maximizes the feasible time the network lasts.

   In order to find out the best link metric for the stated maximization problem, two maximum residual energy path algorithms are presented and simulated.

II. **Maximum lifetime data gathering (MLDA):** MLDA was proposed by Kalpakis et al. models the data routes setup in sensor networks as the maximum lifetime data-gathering problem and presents a polynomial time algorithm.

   The system lifetime depends on the duration for which the schedule remains valid. The aim is to maximize the lifetime of the schedule. The algorithm considers data aggregation while setting up maximum lifetime routes.

III. **Minimum cost forwarding:** Minimum cost forwarding protocol aims at finding the minimum cost path in a large sensor network, which will also be simple and scalable. The protocol is not really flow-based, however since data flows over the minimum cost path and the resources on the nodes are up-dated after each flow, we have included it in this section.

IV. **SAR:** Sequential assignment routing (SAR) is the first protocol for sensor networks that includes the notion of QoS in its routing decisions. It is a table-driven multi-path approach striving to achieve energy efficiency and fault tolerance. The SAR protocol creates trees rooted at one-hop neighbors of the sink by taking QoS metric, energy resource on each path and priority level of each packet into consideration.

   SAR maintains multiple paths from nodes to sink. The limitation is fault-tolerance and easy recovery, the protocol suffers from the overhead of maintaining the tables and states at each sensor node especially when the number of nodes is huge.

V. **Energy-aware QoS routing protocol:** A fairly new QoS aware protocol for sensor networks is proposed by Akkaya and Younis. Real-time traffic is generated by imaging sensors. The proposed protocol extends the routing approach in and finds a least cost and energy efficient path that meets certain end-to-end delay during the connection. The link cost used is a function that captures the nodes energy reserve, transmission energy, error rate and other communication parameters.

VI. **SPEED:** The protocol requires each node to maintain information about its neighbors and uses geographic forwarding to find the paths. In addition, SPEED strive to ensure a certain speed for each packet in the network so that each application can estimate the end-to-end delay for the packets by dividing the distance to the sink by the speed of the packet before making the admission decision. Moreover, SPEED can provide congestion avoidance when the network is congested. SPEED does not consider any further energy metric in its routing protocol. Therefore, for more realistic understanding of SPEED s energy consumption, there is a need for comparing it to a routing protocol, which is energy-aware.

The Table 5.1 summarizes the classification of the protocols that is utilizing data aggregation for energy saving and traffic optimization.

Table 5.1 Classification of routing protocol of wireless sensor network

Routing protocol	Data-centric	Hierarchical	Location-based	QoS	Network-flow	Data aggregation
SPIN	✓					✓
Directed Diffusion	✓					✓
Rumor routing	✓					✓
Shah and Rabacy	✓		✓			
GBR	✓					✓
CADR	✓					
COUGAR	✓					✓
ACQUIRE	✓					
Fe et al.						✓
LEACH					✓	✓
TEEN and APTEEN	✓	✓				✓
PEGASIS		✓				✓
Younis et al.		✓				
Subramanian and Katz		✓	✓			
MECN and SMECH		✓	✓			
GAF		✓	✓			
GEAR		✓	✓			
Chang and Tassiulas		✓			✓	
Kalpakis et al.					✓	
Akkaya et al.		✓	✓	✓		
SAR				✓		
SPEED			✓	✓		

5.6 HIERARCHICAL VS. FLAT TOPOLOGY OUTING

The Flat and hierarchical protocols are different in many aspects. Table 5.2 shows the comparison of the different routing approaches for flat and hierarchical sensor networks.

Table 5.2 Hierarchical vs. Flat topologies routing

Hierarchical routing	Flat routing
Reservation-based scheduling	Contention-based scheduling
Collision avoided	Collision overhead present
Reduced duty cycle due to periodic sleeping	Variable duty cycle by controlling sleep time of node
Data aggregation by clusterhead	Node on multithop path aggregates incoming data from neighbors
Simple but non-optimal routing	Links formed on the fly without synchronization
Overhead of cluster formation throughout the network	Routes formed only in regions that have data for transmission.
Lower latency as multiple hope network formed by clusterheads always available	Latency in waking up intermediate nodes and setting up the multipath
Energy dissipation is uniform	Energy dissipation depends on traffic patterns
Energy dissipation cannot be controlled	Energy dissipation adapts to traffic patterns
Fair channel allocation	Fairness not guaranteed

5.7 NETWORK MANAGEMENT

The network nodes routing from data source to data destination, usually known as sink, is another obstacle in the way of power-aware designing. Node's network need to know in advance the optimal route as per network topology changes dynamically. So, route discovery and tracking management protocols are needed. However, traditional implementations are not feasible in sensor networks because they waste more energy in maintaining routes updated than in transferring information packets.

There are different routing protocols, specific for wireless sensor networks, to reduce the discovery cost. Such routing protocol can be classified into proactive, reactive and hybrid. These protocols are presented below.

I. Proactive protocols, where each node has routing information of every node in the network in storage tables and refreshes these tables periodically or whenever the network changes, such as, Destination-Sequenced Distance Vector (DSDV), Wireless Routing Protocol (WRP), Global State Routing (GSR), Fisheye State Routing (FSR), Source-Tree Adaptive Routing (STAR), Distance Routing Effect Algorithm for Mobility (DREAM), Multimedia support in Mobile Wireless Networks (MMWN), Cluster-head Gateway Switch Routing (CGSR), Hierarchical State Routing (HSR), Optimised Link State Routing (OLSR) and Topology Broadcast Reverse Path Forwarding (TBRPF).

II. Reactive or on-demand protocols were designed to decrease proactive protocols overload and that's why they just keep information about active routes, such as Adaptive On- Demand Distance Vector (AODV), Dynamic Source Routing (DSR), Routing On-demand Acyclic Multi-path (ROAM), Light-weight Mobile Routing (LMR), Temporally Ordered Routing Algorithm (TORA), Associatively-Based Routing (ABR), Signal Stability Adaptive (SSA), Relative Distance Micro-discovery Ad-hoc Routing (RDMAR), Location-Aided Routing (LAR), Ant-colony-based Routing Algorithm (ARA), Flow Oriented Routing Protocol (FORP), Cluster Based Routing Protocol (CBRP), Associatively-Based Multicast (ABAM) On-Demand Multicast Routing Protocol (ODMRP), Adaptative Demand-Driven Multicast protocol (ADMR).

III. Hybrid protocols, whose behavior is both proactive and reactive. This type of protocols divides the network in clusters or trees and uses proactive and reactive methods at different stages: maintain intra-zone routing information refreshed in a proactive fashion whereas inter-zone routes are only updated when are needed. Some of them are: Zone Routing Protocol (ZRP), Zone-based Hierarchical Link State (ZHLS), Scalable Location Update Routing Protocol (SLURP), Distributed Spanning Trees based routing protocol (DST), Distributed Dynamic Routing (DDR), Geographical Routing Algorithm (GRA) and Directed Diffusion Routing.

SUMMARY

Routing in sensor networks is a new area of research, with a limited, but rapidly growing set of research results. In this chapter, presented a comprehensive survey of routing techniques in wireless sensor networks which have been presented in the literature. They have the common objective of trying to extend the lifetime of the sensor network, while not compromising data delivery. Another interesting issue for routing protocols is the consideration of node mobility. Most of the current protocols assume that the sensor nodes and the sink are stationary except in MANET scenario. New routing algorithms are needed in order to handle the overhead of mobility and topology changes in such energy constrained environment.

The chapter also discusses routing protocols. In sensor networks, various forms of routing protocol has attracted a lot of attention in the research field recent years and introduced unique challenges compared to traditional data routing in wired networks. Further discuss possible future research for routing protocols includes the integration of sensor networks with Internet. Most of the applications in security and environmental monitoring require the data collected from the sensor nodes to be transmitted to a server so that further analysis can be done.

QUESTIONS

1. What the requirement of for routing in wireless sensor networks?
2. Broadly explain the classification of wireless sensor networks routing protocol.
3. What are the challenges in routing design?
4. Explain the different routing protocol category.
5. Write a note on Hierarchical routing protocol.
6. What is Location based routing protocols?
7. Brief out the Network flow and QoS-aware protocols.
8. What is network management and classify the routing protocol?

BIBLIOGRAPHY

- Performance Evaluation of Routing Protocols in WSN, Laiali Almazaydeh, Eman Abdelfattah, Manal Al- Bzoor, and Amer Al- Rahayfeh, International Journal of Computer Science and Information Technology, Volume 2, Number 2, April 2010.
- K. Khamforoosh, and H. Khamforoush, "A new rounting Algorithm for Energy Reduction in Wireless Sensor Networks", IEEE, 2009.

- J.N Al-Karaki, and A.E. Kamal, "Routing techniques in wireless sensor networks: a survey", IEEE Wireless Communications, Vol. 11, No. 6, pp.6-28, December 2004.
- M. Younis, M. Youssef and K. Arisha, "Energy-Aware Routing in Cluster-Based Sensor Networks", in the Proceedings of the 10th IEEE/ACM(MASCOTS2002), Fort Worth, TX, October 2002.
- R. V. Biradar, V. C. Patil, Dr. S. Sawant, and Dr. R. R. Mudholkar, "Classification and comparison of routing protocols in wireless sensor networks", UbiCC Journal, Vol. 4.
- S. R. Das, C. E. Perkins, and E. M. Royer. Performance comparison of two on-demand routing protocols for ad hoc networks. In Proceedings of the IEEE Infocom, pages 3-12, Tel Aviv, Israel, March 2000.
- S. Singh, M. Woo, and C. Raghavendra. Power aware routing in mobile ad hoc networks. In Proceedings of the ACM/IEEE International Conference on Mobile Computing and Networking, pages 181-190, October, 1998.
- M. Abol, T. Wysocki, and E. Dutkiewicz. "A review of routing protocols for mobile ad hoc networks". Ad Hoc Networks, 2(1), January 2004.
- M. Ilyas, Ed., The Handbook of Ad-Hoc Wireless Networks, 1st ed., ser. The Electrical Engineering Handbook Series. CRC Press, 2003.
- K. Akkaya and M. Younis, "A Survey of Routing Protocols in Wireless Sensor Networks, " in the Elsevier Ad Hoc Network Journal, Vol. 3/3 pp. 325-349, 2005.
- S. Servetto and G. Barrenechea\Constrained Random Walks on Random Graphs: Routing Algorithms for Large Scale Wireless Sensor Networks", proceedings of the 1st ACM International Workshop on Wireless Sensor Networks and Applications, Atlanta, Georgia, USA, 2002.

6

DATA AGGREGATION IN WSN

6.1 INTRODUCTION

Wireless sensor networks (WSNs) are increasingly being used to monitor various parameters in a wide range of environmental monitoring applications is a low power and small size sensor nodes. Recent years have witnessed a growing interest in the application of wireless sensor networks in unattended environments. The event detection using wireless sensor network could be an device crossing the line of control, rise of ambient temperature or pressure above some threshold, camera detecting a burglar in a sensitive area, disaster control applications like detection and prediction of Landslide, Volcano, Tsunami or Earthquake etc. Other examples include target tracking, industry monitoring, precision agriculture etc.

In many instances, environmental scientists are interested in collecting raw data using long-running queries injected into a WSN for analyzing at a later stage, rather than injecting snap-shot queries containing data-reducing operators (*e.g.*, MIN, MAX, AVG) that aggregate data. Collection of raw data poses a challenge to WSNs as very large amounts of data need to be transported through the network. This not only leads to high levels of energy consumption and thus diminished network lifetime but also results in poor data quality as much of the data may be lost due to the limited bandwidth of present-day sensor nodes. Ubiquitous Computing envisions the era of computing such computing services are available to all anywhere, any time. In order to provide such pervasive services, a ubiquitous system collects a large amount of information continuously from its physical and computational environment. The information is manage and manipulate optimally is one of the most important aspects of ubiquitous computing system.

From the survey paper of I F Akylidiz (2002) sensor node spent more energy for communication than for processing. Hence, there is a need for efficient data gathering algorithms to increase the lifetime of sensor network. These sensors can then be deployed in large numbers to self-organize into networks that serve a wide range of purposes. The laid is for many-to-one type of communication or many-to-few communication modes for data-gathering architecture followed in wireless sensor network. Communications within each cluster are again of the many-to-one type, *i.e.*, data flows from each sensor to the cluster head, where they can be processed, compressed, aggregated and relayed. More broadly, clustering organize data collection communications in a large-scale network. The possible organizations of the network include the flat and hierarchical organizations. In a flat organization all nodes/sensors act as peers in transmitting and relaying data for one another. In a hierarchical network, layers of clusters are formed. Nodes send their data to the cluster heads who then relay the data to either a higher layer cluster head or the sink.

6.2 DATA AGGREGATION

Data Aggregation, derived from the aggregation of two or more contributing data characteristics. Data aggregation has been put forward as an essential paradigm for wireless routing in sensor networks. The idea is to combine the data coming from different sources route to control over eliminating redundancy, minimizing the number of transmissions and thus saving energy.

An important energy saving mechanism for sensor nodes is to exploit in-network data aggregation. In wireless sensor networks the raw sensed data is typically forwarded to a sink (gateway) node for processing. The main idea of in-network data aggregation is to eliminate unnecessary packet transmission by filtering out redundant sensor data and/or by performing an incremental assessment of the semantic of the data.

Aggregation can be made from different data occurrences within the same data subject and a de-normalized database. However, since various sensor nodes often detect common phenomena, there is likely to be some redundancy in the data the various sources communicate to a particular sink. Nodes in such applications are equipped with limited energy supply and need careful management in order to extend their lifetime. In order to conserve energy, many of the routing protocols proposed for wireless sensor networks reduce the number of transmitted packets by pursuing in-network data aggregation. Almost, all of the aggregation is to save sensor's energy while considering unconstrained data traffic. There are efficient algorithms for achieving maximal possible energy saving during data aggregation.

The simplest and most commonly used form of summarization is aggregation that represents the collection of information in fewer values and uses this aggregate information to reply the queries later. The simplest form of aggregation is average or mean. However, the context type and based on application requirement, the more suitable forms are variance, standard deviation and rate of change. For example, as shown in Table 6.1 if a temperature sensor is giving temperature values after every 5 minutes, then we can repeatedly calculate and generate aggregates.

Table 6.1 Aggregate values of temperature

Date	Period	Avg. Temp	Min. Temp	Max. Temp
09/18	Morning	16	13	18
09/18	Afternoon	20	18	23
09/18	Evening	18	17	20
09/19	Morning	17	14	18

Recent research on data aggregation in sensor networks focused on generating optimal aggregation trees for reduced energy consumption. The proposed mechanisms promote path sharing as much as possible and therefore trade increased per node queuing delay, and consequently boost the overall delivery latency, for further energy savings. Moreover, significant attention has also been dedicated by the database community to the development of lightweight query languages and tool suite that enables task level analysis of the potential aggregation.

6.3 SENSOR NETWORK ARCHITECTURE

Wireless sensors network spread throughout an area of interest to monitor the possible events in the area. The sensors are battery-operated with diverse capabilities and are empowered with limited data processing engines. These sensors are dynamically changing nodes that are stationary or mobile. A sensor network consists of one or more "sinks" which subscribe to specific data streams by expressing interests or queries. The sensors in the network act as "sources" which detect environmental events and push relevant data to the appropriate subscriber sinks. An important energy conservation technique for WSNs is to approximate the time series captured by a sensor and synchronize the approximation with the sink.

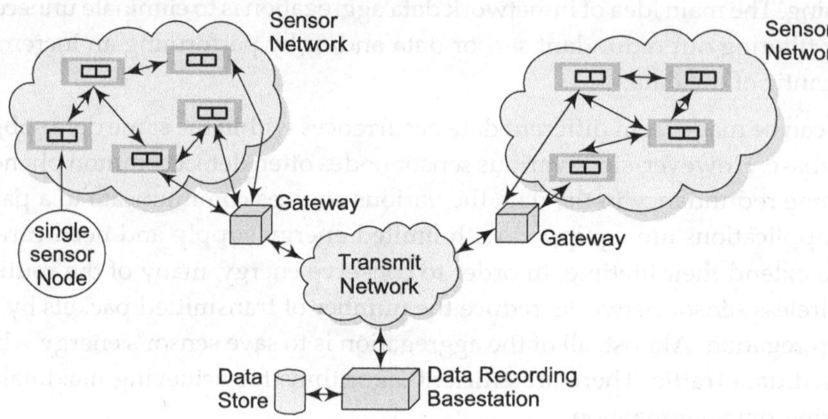

Fig. 6.1 System architecture for wireless sensor networks (WSN)

Wireless sensor networks architecture is depicted in Fig. 6.1 pose unique challenges with regards to power consumption, and overall size, so the security protocols used for sensor data protection must be efficient, resource friendly and fast. A gateway node is a less energy-constrained node deployed in the physical proximity of sensors. The gateway is responsible for organizing the activities at sensor nodes to achieve a mission, fusing data collected by sensor nodes, coordinating communication among sensor nodes and interacting with command nodes via transient network. The gateway node sends to the command node reports generated through fusion of sensor readings, e.g. tracks of detected targets. The command node presents these reports to the user and performs system level fusion of the collected reports for overall situation awareness.

All the sensors are assumed to be within the communication range of the gateway node The application architecture of sensor network nodes is shown in the Fig. 6.2. The sensor is assumed to be capable of operating in an active mode or a low-power stand-by mode.

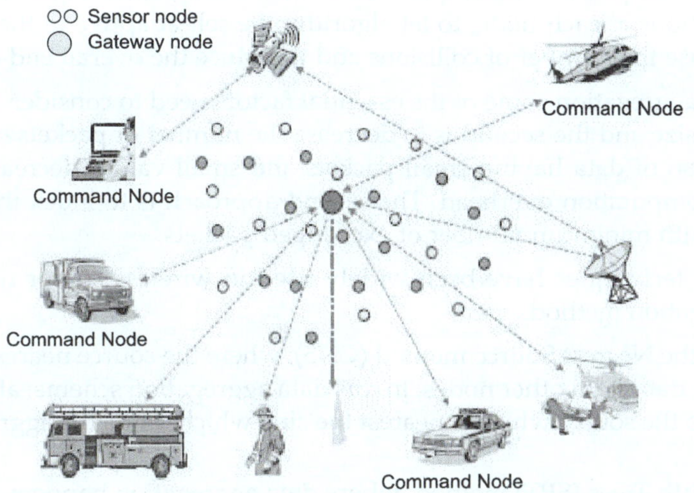

Fig. 6.2 The application architecture of sensor network

The sensing and processing circuits can be powered on and off. In addition the radio transmitter and receiver can be turned on and off and the transmission power can be programmed for a required range. It is also assumed that the sensor can act as a relay to forward data from another sensor.

Moreover, a sensor is assumed to switch between generating real time and non-real-time data.

6.4 DESIGN AREAS ON DATA AGGREGATION

Some of the design criteria for the data aggregation are as follow:

I. Data aggregation for sensor networks exploits latency due to hop sequence leading to aggregating and sending its data for the next hop. Node waiting too much can increase the end-to-end delay. So, an algorithm shall aggregate tree for real-time packets and strives to meet the required bound on the end-to-end delay when two types of traffic coexist in the network.

II. Another in-network data aggregation scheme aims at minimizing the end-to-end delay by concatenating multiple packets into one at the MAC layer. The idea is to limit the medium access contention so that the packet queuing delay will be reduced. MAC layer and optimization at network layer is to consider otherwise when the concatenated packet is dropped then the recovery process will be very expensive, diminishing the energy and latency gains.

III. The optimal aggregation tree is modeled as a minimum Steiner tree NP-hard problem. Here, data is sent along the Shortest Path Trees (SPT), where source to gateway aggregated at common intermediate hops use the SPT heuristic in approach to build an initial aggregation tree.

6.5 DATA AGGREGATION TECHNIQUE

Many researchers tackled the data aggregation techniques in wireless sensor networks including query processing and data handling. There has been a lot of work and approaches on query processing in distributed database systems took place. Sensor networks have very limited power, small memory computational power and limited bandwidth. Some of the requirements in sensor networks are to control the power consumption in a data management algorithm, to decrease the number of computations at each node, to let algorithm be self-adaptive to the changing network conditions, to decrease the number of collisions and to reduce the overall end-end latency.

To optimize data aggregation some of the essential factors need to consider. First approach is to decrease the packet size and the second is to decrease the number of packets sent. In large sensor networks, aggregation of data having small packets and small values decreases the power consumption and the computation overhead. The second approach is to index the network so to be able to query data with minimum number of exchanged packets.

Data aggregation techniques have been widely used in wireless sensor networks. The basic forms of data aggregation methods are:

I. The Center at the Nearest Source method (CNS), where the source nearest to the destination aggregates the data from other nodes. In this data aggregation scheme, all sources send their data directly to the source which is nearest the sink which sends the aggregated information on to the sink.

II. The shortest Path Trees (SPT) method, where data aggregation happens at the intermediate nodes within a shortest path tree rooted at the sink. In this data aggregation scheme, each source sends its information to the sink along the shortest path between the two, and overlapping paths are combined to form the aggregation tree.

III. The greedy Incremental Trees (GIT) method, where an aggregation tree is constructed by connecting each destination sequentially to the existing tree via a shortest path. In this scheme at the aggregation tree consider the process of forming the shortest path between the sink and the nearest source.

Aggregation dramatically reduces the amount of data routed through the network, increasing throughput and extending the lifetime of battery powered sensor networks.

6.5.1 Data Propagation and Aggregation

Data propagation is expressed based on the directed network tree given in Fig. 6.3 with two message sources, one at Node A and the other at Node B. Network message transmission is started at the lowest network level, *i.e.*, at the node furthest away from the sink node and proceeds on a level by level basis toward the sink node. It assigns a relevant strength of significance to the information that each node is transmitting into the network.

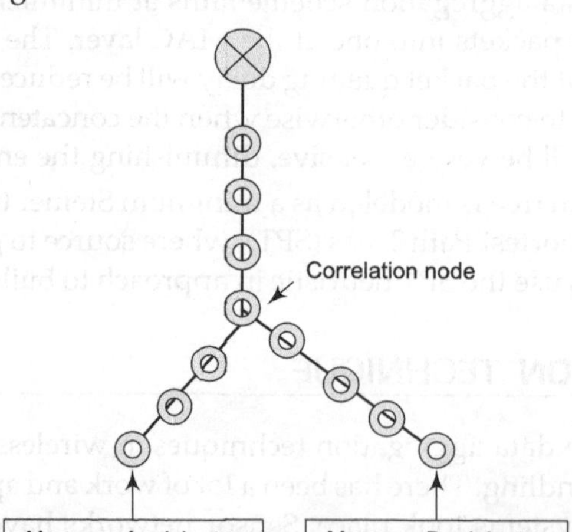

Fig. 6.3 Simple network structure

Further a node gathers the data, assembles the appropriate data message, and checks the relevant entry in the interest cache to retrieve the ID of the neighboring node. Before transmitting an incoming message to its neighbors, a node checks its data cache to see if there are any existing recent data messages that contain data. Using this knowledge route the messages to certain specific nodes in more energy efficient way.

As shown in Fig. 6.4 the layered data aggregation, the transmission of data messages being sent back from the various source nodes across the network finally leads to the formation of data aggregation points.

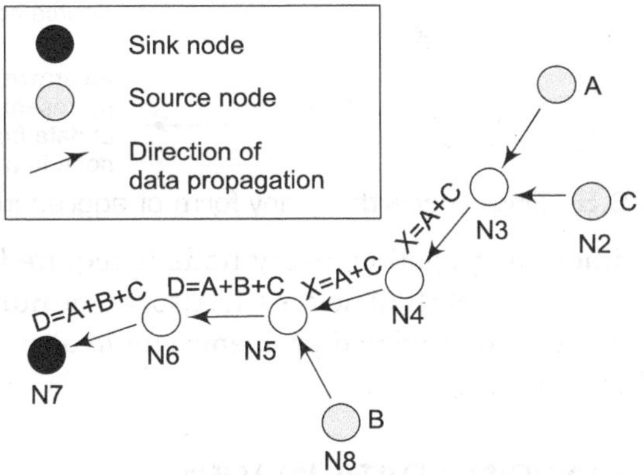

Fig. 6.4 Layered aggregations

Here, the message is sent out by a sink node containing a query, D which has the following definition, D = A + B + C. The parameters A, B, and C are the fundamental components that can be understood by certain nodes in the sensor network that are capable of providing the relevant data. The two nodes, N1 and N2, capable of providing data for parameters A and C which are geographically close to one another. The node N5 receives data for parameters A and C.

All nodes shown above belong to different clusters allows data to be actually processed within the network and subsequently aggregated thus resulting in the reduction of data as close to the source nodes as possible.

6.6 BENEFITS OF DATA AGGREGATION

There are generally two factors to consider for analyzing the benefits for performing data aggregation. It is the total number of messages transmitted by every individual node and to the size of a message received.

The most significant savings can be obtained when using the partial aggregation method involving either distributive or algebraic operators. In such an instance, for every round of a query, each node transmits a maximum of only one message regardless of its hierarchy in the tree. Algebraic aggregates require message sizes that are slightly larger than holistic aggregates. For instance when calculating an average, the data message needs to contain not just the SUM but the COUNT as well. If the average operation was carried out using holistic aggregation however, a data message would only contain the value generated by a certain source node.

In holistic the message size is slightly smaller, every intermediate node is required to retransmit every message it has received from its child nodes. Thus, the larger number of child nodes will have, the larger the number of messages transmitted. Effectively, this implies that source nodes which lie much lower in the hierarchy consume a lot less energy than their counterparts which are positioned closer to the sink node.

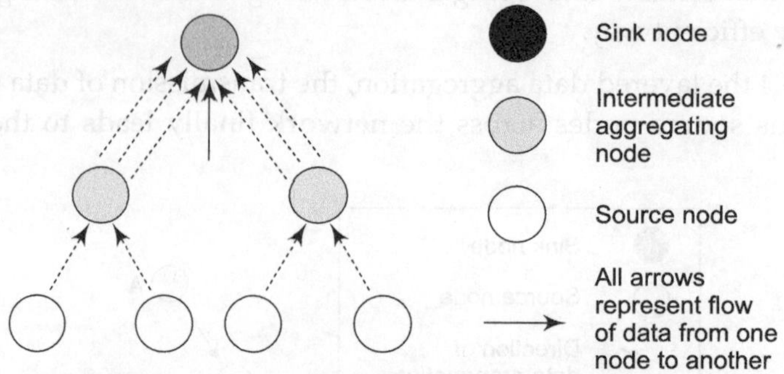

Fig. 6.5 Data flow without any form of aggregation

Fig. 6.4 illustrates how without aggregation, every node is required to transmit n times more number of packets, where, n = Total no: of child nodes + 1. Also, as the number of messages increases significantly closer to the sink node, other related problems would also arise, *e.g.* traffic congestion, bandwidth limitations and increased latency.

6.7 DESIGNS FOR ENHANCING DATA IN WSN

Periodic collection of aggregated data from sensors to a common sink over a tree topology is a fundamental operation in wireless sensor networks. Many WSN applications require periodic summaries or aggregates of the data rather than raw sensor readings. In such cases, data coming from different sources can be aggregated at each hop en-route to the sink.

The fastest rate at which data can collect from a stream of nodes has its impotence. The "aggregated" data of wireless sensors organized is a hierarchy technique by considering TDMA scheduling. In the TDMA, time slots number is minimized to schedule each link of the data aggregation tree. The other technique is to combine the scheduling with transmission power control to reduce the effects of interference.

In many WSN applications, it is of interest to maximize the rate at which the sink can receive aggregated data from the network.

6.7.1 The Preliminary Design Details and Assumptions

In the following Fig. 6.6 shows the relationship between the schedule length and the aggregated data rate. Here, the numbers on the links shown in the assigned time slots and the numbers inside the circles represent the node ids. On the left of the figure it shows the schedule of the received packets from the associated senders by each parent on each time slot. After frame 1, once the sink gets initial data from each source a pipeline is established, the same schedule is repeated and the sink collects the aggregated data from the network at a rate of 3 time slots. Thus, the schedule length should be minimized to improve the data collection rate.

	Frame 1			Frame 2		
Receiver	slot 1	slot 2	slot 3	slot 1	slot 2	slot 3
Sink	3	1.4	2, 5, 6	3, 7	1, 4	2, 5, 6
Node 1	4			4		
Node 2	5	6		5	6	
Node 3		7			7	

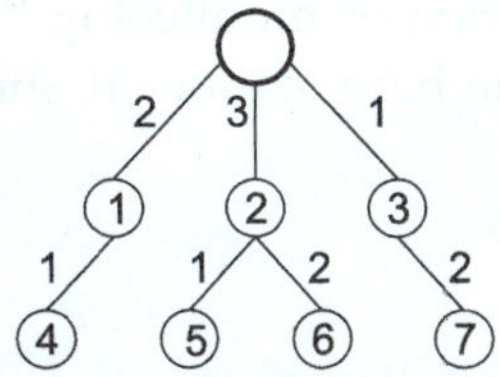

Fig. 6.6 Relationship between data collection rate and schedule length

Here, time is divided into equal sized slots that are grouped into frames to minimize the length of the frame such that each node is assigned one time slot. Time slot assignment is based on the following constraints: 1. no node can transmit and receive at the same time slot, 2. no node can receive from more than one transmitter at a time slot. If a receiver experiences excessive levels of interference due to simultaneous transmissions in a time slot, the transmitter should not be transmitting in that time slot.

To assume all the nodes in the network are sources and the data is aggregated such that the data coming from different sources are combined into a packet(s) before forwarding. In the static WSN the sensor nodes periodically sense the environment and send their readings over a multi-hop tree topology to a sink node. In the minimum-hop routing trees all the nodes select a parent node to transmit their readings to be forwarded towards the sink node.

If the incoming packets cannot be combined in a single packet then multiple packets have to be forwarded, This is a reasonable assumption the size of the sensor packet is small and each time slot is long enough to transmit those packets.

6.7.2 System Design for Data Aggregation

Protecting the data privacy in many wireless sensor network applications is a major concern. The following criteria summarize the desirable characteristics of a private data aggregation scheme:

Privacy: Each nodes data should be only known to itself. The data aggregation scheme should be able to handle attacks and collusion among compromised nodes. When a sensor network is under a malicious attack, it is possible that some nodes may collude to uncover the private data of other nodes. A good private data aggregation scheme should be robust to such attacks.

Efficiency: The goal of data aggregation is to reduce the overhead *i.e.*, the number of messages transmitted within the sensor network, thus reduce resource and power usage. Data aggregation achieves bandwidth efficiency by using in-network processing. However, a good private data aggregation scheme should keep that overhead as small as possible.

Accuracy: An accurate aggregation of sensor data is desired, with the constraint that no other sensors should know the exact value of any individual sensor. Accuracy should be a criterion to estimate the performance of private data aggregation schemes.

6.7.3 The Hierarchical Network Architecture

The hierarchical network architecture is shown in Fig. 6.7, consisting of gateway nodes, normal sensor nodes, and the sink node. In this application, the gateway nodes were implemented with an embedded Linux system supporting IEEE 802.11b and were more powerful than normal sensor nodes. It is to reduce the energy consumption on the gateway nodes since the energy depletion of a gateway node will disrupt the communication of all sensor nodes within its cluster, resulting in network partition or monitoring "black holes," from which measurement information cannot be sent out to the base station. It shows the value of sensor clustering, scheduling, and message aggregation.

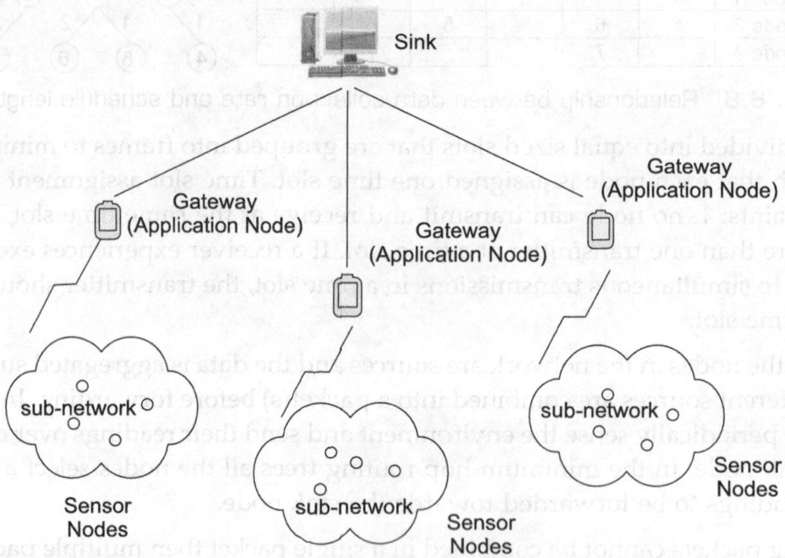

Fig. 6.7 Hierarchical architecture of sensor networks

In a tree system structure the lowest layer, sensor nodes deployed in dense patches that were widely separated. Each sensor patch included a gateway node responsible for local data collection, data aggregation, and long-distance radio transmission directly to a remote base station (also the sink), which was connected to database over the Internet.

6.8 ACO FOR DATA AGGREGATION

Ant colony optimization system (ACO), a population-based algorithm, provides natural and intrinsic way of exploration of search space in optimization settings in determining optimal data aggregation. Data aggregation is important in energy efficient routing in energy constraint wireless sensor networks which exploits correlated sensing data and aggregates at the intermediate nodes to reduce the number of messages exchanged network. The complexity of optimal data aggregation is NP-hard.

In data gathering application large amount of communication network aggregation is to achieve maximum lifetime of network. In-network data aggregation is an important in energy constraint sensor network that exploits sensing data and aggregates at the intermediate nodes to reduce the number of message in the network.

Active research in area of sensor network aims for finding efficient approximation algorithms for optimal aggregation problem. Optimal aggregation is modeled as combinatorial optimization problem which is solved using population based metaheuristic approach Ant Colony Optimization (ACO).

6.8.1 Ant Colony Optimization for Optimal Aggregation Tree

The Ant Colony Optimization algorithm runs in two methods.

I. In the forward algorithm, the route is constructed by one of the ants in which other ants search the nearest point of previous discovered route. The points, where multiple ants join are aggregation nodes.

II. In the backward algorithm pass nodes of the discovered path are given weight in form of node potential which indicates heuristics for reaching to destination node or nearest aggregation node and pheromone trails is the heuristics to communicate other ants of the route discovered.

Ants tries to follow the route to get pheromone eventually converges to the optimal route. Non-optimal route pheromone gets evaporated with time. The aggregation points on the optimal tree identify data aggregation. The indicator in data aggregation points gives estimate of number of paths aggregates in it.

6.8.2 Heuristics for Global Optimization of Aggregation Tree

In large sensor network, finding optimal aggregation tree is NP-Hard problem. The reduction is weighted set covering problem. In the Ant-aggregation algorithm, source nodes constructs local best aggregation tree associated with a cost. The more weight in cost function will converge in optimal aggregation points. The algorithm iterates to search the global best and the convergence of optimal aggregation tree algorithm constructed by ant routing in iterations.

SUMMARY

In this chapter discussed the data aggregation, aggregate queries in a wireless sensor networks. Wireless sensor networks are an important type of resource-constrained distributed event-based system. The formation of an optimal data aggregation tree is generally NP-hard.

Furthermore as for future work in the wireless sensor network as the work is getting more and more dense, approach should confront with the difficulties of topology construction, data routing, loss tolerance by including several optimization techniques that further decrease message costs and improve tolerance to failure and loss.

The data aggregation tree generation is heuristics. It focuses for effective aggregation of the extracted data.

QUESTIONS

1. What is meant by data aggregation?
2. What are the design criteria for the data aggregation?
3. Draw and explain the System Architecture for Wireless Sensor Networks.
4. Discuss the data propagation and aggregation technique.
5. List out the benefits of data aggregation.
6. What is the soft technique for data aggregation?

BIBLIOGRAPHY

- T. He, B. M. Blum, J. A. Stankovic, and T. F. Abdelzaher, "Aida: Adaptive application independent data aggregation in wireless sensor networks," ACM Transactions on Embedded Computing Systems, (to appear) 2004.
- C. Intanagonwiwat, D. Estrin, R. Govindan, and J. Heidemann, "Impact of network density on data aggregation in wireless sensor networks", In ICDCS-22, November 2001.
- B. G. Jagyasi, B. K. Dey, S. N. Merchant, and U. B. Desai, .An mmse based weighted aggregation scheme for event detection using wireless sensor network,. 14th European Signal Processing Conference, EUSIPCO 2006, Sep 2006.
- Efficient Gathering of Correlated Data in Sensor Networks, Himanshu Gupta, Vishnu Navda, Samir Das, and Vishal Chowdhary, State University of New York, Stony Brook.
- An Efficient Data Gathering Scheme for Wireless Sensor Networks, C.Tharini, P. Vanaja Ranjan, European Journal of Scientific Research, ISSN 1450-216X Vol.43 No.1 (2010).
- Myung Ho, Yeo Mi, Sook Lee Seok, Jal Lee Jue, Soo Yoo, 2008. "Data Correlation -Based clustering in Sensor Networks", International Symposium on Computer Science and its Applications IEEE 2008.
- Chong Liu, KuWu, Jian Pei, 2007 "An Energy efficient Data Collection Framework for Wireless Sensor Network by exploiting spatiotemporal Correlation" IEEE Transactions on Parallel and Distributed Systems Vol. 18, No.7
- Data-gathering wireless sensor networks: organization and capacity, Enrique J. Duarte-Melo, Mingyan Liu, 2003 Elsevier.
- B. Krishnamachari, D. Estrin and S. Wicker, "Impact of Data Aggregation in Wireless Sensor Networks," in Proceedings of the 22nd International Conference on Distributed Computing Systems Workshops (ICDCSW02), pp. 575-578, 2002.
- Towards Using Data Aggregation Techniques in Ubiquitous Computing Environments, Faraz Rasheed, Young-Koo Lee, Sungyoung Lee, 2006 IEEE Computer Society.
- Modeling Data Gathering in Wireless Sensor Networks, Bhaskar Krishnamachari, and 2005 Springer.
- M. Varshnery and R. Bagrodia, Detailed models for sensor network simulations and their impact on network performance, MSWim'04, October, 2004, Venezia, Italy.
- Annamalai, S. K. S. Gupta, and L. Schwiebert, On tree-based converge casting in wireless sensor networks, WCNC 2003-IEEE Wireless Communication and Networking Conference, 4(1) (2003), 1942-1947.
- P. Gupta and P. R. Kumar, "The capacity of wireless networks," IEEE Transactions on Information Theory, vol. 46, no. 2, pp. 388-404, March 2000.

7

POWER MANAGEMENT IN WSN

7.1 INTRODUCTION

Wireless Sensor Networks (WSNs) are large scale networks of sensor nodes. The numbers of wireless nodes communicate with the thousands of separate devices including sensor equipment for data collection and control actions. A wireless sensor node is a good example for a System on Chip (SoC) that has communication, computation, sensing and storage capabilities. These miniaturized nodes have stringent constraints in terms of available resources like processing power, battery power, program memory, available bandwidth. Fig. 7.1 shows schematic diagram of sensor node components. Basically, each node comprises of a microcontroller, power source, Radio Frequency (RF) transceiver, external memory, and sensors. These sensor nodes are used in wide variety of applications nowadays.

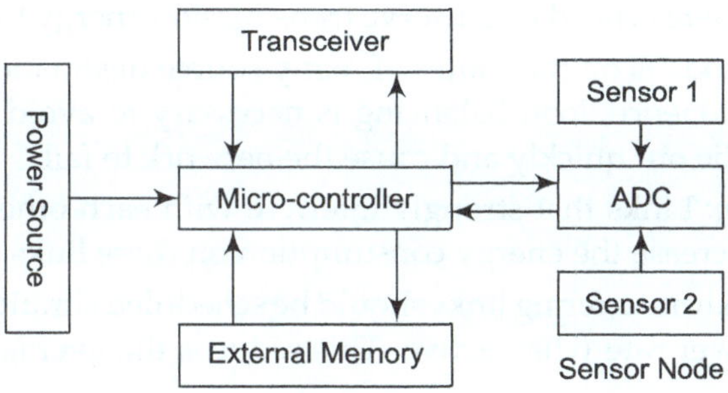

Fig. 7.1 Sensor node architecture

A WSN typically consists of hundreds or thousands of sensor nodes. These nodes have the capability to communicate with each other using multi-hop communication. Typical applications of these WSN include but not limited to monitoring, tracking, and controlling. The basic functionality of an operating system is to hide the low-level details of the sensor node by providing a clear interface to the external world. Processor management, memory management, device management, scheduling policies, multi-threading, and multitasking are the low level services be provided by an operating system. In addition suitable operating system is required for WSN to provide these functionalities to facilitate the user in writing applications easily with little knowledge of the low-level hardware. Fig. 7.2 depicts, where operating system stands in the software layers of the WSN. Middleware and application layers are distributed across the nodes. Core kernel of the operating system sits at each individual node. On top of it, middleware and applications run as interacting modules across nodes.

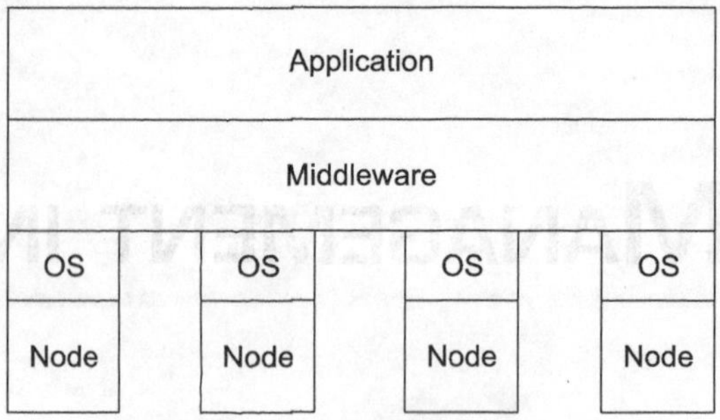

Fig. 7.2 Software layers

Wireless Sensor Networks have gained increasing attention from both the research community and actual users. The existing research on energy consumption of sensors is usually based on either theoretical models or computer simulations. The sensor nodes are generally battery-powered that face concerns on how to reduce the energy consumption to increase the network lifetime that can be extended to reasonable times.

7.2 POWER MANAGEMENT

A typical constraint for wireless sensor networks is power. For energy-constrained wireless networks, the network lifetime by using transmission schemes shall have the following characteristics.

I. *Multi-hop routing:* In wireless environments the received power typically falls off as the power of distance. Hence, there is need to conserve transmission energy by using multihop routing.
II. *Load Balancing*: If a node is on the routes of many source destination pairs, it will run out of energy very quickly. Hence, load balancing is necessary to avoid the creation of hot spots where some nodes die out quickly and cause the network to fail.
III. *Interference mitigation:* Links that strongly interfere with each other should be scheduled at different times to decrease the energy consumption on these links.
IV. *Frequency reuse*: Weakly interfering links should be scheduled simultaneously so that each link can transmit at a lower rate when active. This reduces the average transmission power on each link.

Most wireless sensor nodes operate with non-renewable batteries. Given the potential inaccessibility of the deployment setting, it may not be possible to change batteries or to recharge them. In many scenarios, wireless sensor nodes may not be reused once they exhaust their energy.

In general when fully active, a sensor node requires from 1 to 50 mW. A power management policy can be applied to different components of a sensor node at different circumstances in order to minimize these requirements. In general, power is managed by driving the following policies:

(*i*) Sleep (memory standby, interrupts active, clocks active, CPU off);

(*ii*) Sleep (memory retained, interrupts active, clocks active, CPU off); and,

(*iii*) Sleep (memory retained, interrupts active, clocks off, CPU off).

Due to a deliberate node redundancy, the lifetime of a wireless sensor network can be optimized by making most of the sensor nodes sleep while a selected few act as sentinels to awake the others when an interesting event is emerging. Depending on the urgency of capturing an event, any one of the above policies can be applied to minimize power consumption.

7.3 FACTORS OF ENERGY CONTROL

Wireless sensor networks consist of a distributed set of sensor a node, each of which senses, computes, and communicates with each other to cooperatively work on a set of tasks. These sensor nodes typically use battery operated sensing and computing devices and a low-power radio transmitter and receiver. Effective integration of computation and communication in sensor networks is a challenge to solve that will contribute to parallel and distributed computing, *e.g.*: dynamic network topology, computational power constraints, self-configurability, and limited memory.

Typically, a sensor node is a tiny device that includes three basic components:

(*i*) A sensing subsystem for data acquisition from the physical surrounding environment,

(*ii*) Processing subsystem for local data processing and storage, and

(*iii*) A wireless communication subsystem for data transmission.

This power source often consists of a battery with a limited energy budget. It is impossible or inconvenient to recharge the battery, because nodes may be deployed in a hostile or unpractical environment. On the other hand, the sensor network should have a lifetime long enough to fulfill the application requirements by prolonging the lifetime of the battery. Hence, the energy conservation is a key issue in the design of systems based on wireless sensor networks.

7.4 REASONS FOR ENERGY WASTE

Because of various limitations and the characteristics of wireless sensor networks, the low power consumption is the main criterion for protocol design at every layer. The medium access control layer is one of the interesting research areas, and provides large opportunities of energy savings by dealing with the situations among nodes.

There are several major sources of energy waste in wireless sensor networks:

(*i*) Collision occurs when two nodes transmit at the same time and interfere with each other's transmission. Hence, re-transmissions increase energy consumption.

(*ii*) Control packet overhead such as RTS/CTS/ACK can be significant for wireless sensor networks that use small data packets.

(*iii*) Overhearing means that there is no meaningful activity when nodes receive packets or a part of packets that are destined to other nodes.

(*iv*) Idle listening is the cost of actively listening for potential packets. Because, nodes must keep their radio in receive mode, this source causes inefficient use of energy.

The sensor network model shown in Fig. 7.3 consisting of one sink node (or base station) and a large number of sensor nodes deployed over a large geographic area (sensing field). Data are transferred from sensor nodes to the sink through a multi-hop communication paradigm.

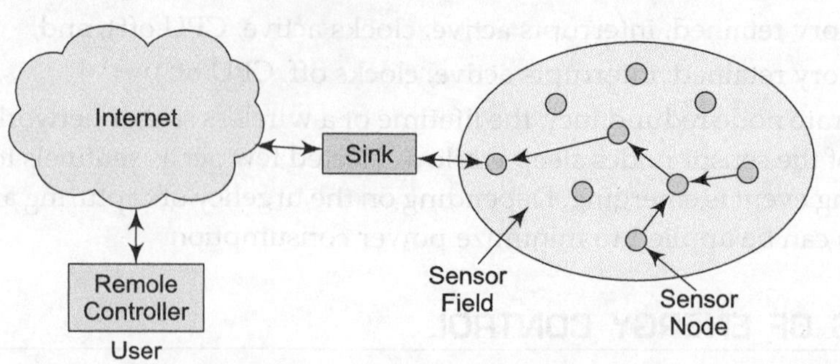

Fig. 7.3 Sensor network architecture

The sink and the sensor nodes may be dynamic or static sensor network. Generally, data transmission is very expensive in terms of energy consumption, while data processing consumes significantly less. For transmitting a single bit of information the energy cost is approximately the same as that needed for processing a thousand operations in a typical sensor node. Power management schemes are used for switching off node components that are not temporarily needed. The energy efficient protocols are aimed at minimizing the energy consumption during network activities. However, a large amount of energy is consumed by node components such as CPU, radio, etc.

The wireless communication often dominates the energy dissipation in a WSN, and several promising approaches have been proposed to achieve power-efficient multihop communication in ad hoc networks. There is the need for low cost devices so some of the most important challenges in WSNs are related to energy-efficiency, scalability, routing, mobility, reliability, timeliness, security, clustering, localization and synchronization.

However, to become a reality, many new problems and challenges overcome in WSNs as their paradigm differs from traditional wireless networks thereby a new protocol is evolved. With an inexpensive hardware it is possible to build a distributed, self-configured network of adaptive sensors which can be used to monitor dangerous or inhospitable environments. Unfortunately, one of the main limitations of these platforms is the limited operational lifetime due to high power consumption.

7.5 ENERGY MANAGEMENT IN WIRELESS SENSOR NETWORK

The energy efficiency of a node is defined as the ratio of the amount of data delivered by the node total energy expanded. Higher energy efficiency implies that a greater number of packets can be transmitted by the node with a given amount of energy reserve. The main reasons for energy management in wireless sensor network are listed below:

(*i*) *Limited energy reserve*: The main reason for the development of ad-hoc wireless network is to provide a communicative infrastructure in the environments where the setting up of infrastructure is impossible. Advance in battery technologies have been negligible as compared to the recent advance that have taken up in the fields of mobile computing and communication. The increase gap between power consumption requirement and power availability adds to the importance of energy management.

(*ii*) *Idle listening*: is the major power consumption source for many networks. For most transceivers, the receive mode power consumption is on the same order of magnitude as the transmission power, and most MAC protocols put the transceiver in receive mode, whenever it does not transmit, whether there is the need to receive a message or not.

(*iii*) *Difficulties in replacing the batteries:* The sensor nodes that are deployed in the remote areas consist of batteries. It is something very difficult to replace or recharge the batteries in a situation like battlefields etc.

(*iv*) *Retransmissions:* resulting from collisions can be quite significant if the network load is high and the collisions frequent.

(*v*) *Lack of central coordination:* The lack of a central coordinator, such as the base station in cellular networks, introduces multi-hop routing and necessitates that some of the intermediate nodes act as relay nodes. If the proportion of relay traffic place an important role in ad hoc wireless networks is large, than it may lead to a faster depletion of the power source for that node. On the other hand, if no relay traffic is allowed through a node, it may lead to partitioning of the network.

(*vi*) *Control packet:* overhead (*e.g.*, RTS, CTS, ACK) can be significant for sensor networks which, typically, have small packets.

(*vii*) *Constraints on the battery source:* Due to batteries increase the size and weight of a mobile node its reduction in the size of the battery results in less capacity which, in turn, decreases the active lifespan of the node. Hence, reducing the size of the battery, it is to formulate energy management techniques to utilize the battery capacity in the best possible way.

(*viii*) *Selection of optimal transmission power:* The consumption of battery charge increases with an increase in the transmission power. The transmission power selected determines the reach ability of the nodes. An optimal value for the transmission power decreases the interference among nodes, which, in turn, increases the number of simultaneous transmission.

(*ix*) *Transmitting power:* Unnecessarily high transmitting power not only results in higher power consumption, but may also increase the interference at other nodes in the network.

(*x*) *Channel utilization:* A reduction in the transmission power increase frequency reuse, which lead to better channel reuse. Power control become very important and maintains the signal-to-interference ratio (SIR) at the receiver to increase channel reusability. Sub-optimal utilization of the available resources will save the channel utilization.

7.6 ENERGY WASTE IN MAC PROTOCOL

The major sources of energy waste in a MAC protocol are the following:

(*i*) *Idle listening*: Idle listening is the major power consumption source for many networks. For most transceivers, the receive mode power consumption is on the same order of magnitude as the transmission power, and most MAC protocols put the transceiver in receive mode whenever it does not transmit, whether there is the need to receive a message or not.

(*ii*) *Collision:* When a transmitted packet is corrupted it has to be discarded, and the follow-on retransmissions increase energy consumption. Retransmissions resulting from collisions can be quite significant if the network load is high and the collisions frequent.

(*iii*) *Control Packet Overhead*: (*e.g.*, RTS, CTS, ACK) can be significant for sensor networks which, typically, have small packets. While Sending and receiving control packets it consumes energy too.

(*iv*) *Overhearing*: meaning that a node picks up packets that are destined to other nodes.

Unnecessarily high transmitting power not only results in higher power consumption, but may also increase the interference at other nodes in the network.

7.7 ENERGY AWARE ROUTING FOR CLUSTER BASED SENSOR NETWORK

Younis et al.proposed a hierarchical routing algorithm based on a three tier architecture. Sensors are grouped into clusters prior to network operation. The algorithm utilize cluster heads, namely gateways, which are less energy constrained than sensors. These sensor nodes know the location. Gateways preserve the states of the sensors and sets up multi-hop routes for collecting sensors data. A TDMA based MAC protocol is used for nodes to send data to the gateway. The gateway informs each node about slots in which it should listen to other nodes transmission and slots. The command node (sink) communicates only with the gateways.

The sensor is assumed to be capable of operating in an active mode or a low-power stand-by mode. The sensing and processing circuits can be powered on and off. In addition both the radio transmitter and receiver can be independently turned on and off and the transmission power can be programmed based on the required range. The sensor nodes in a cluster can be in one of four main states:

(*i*) Sensing only,
(*ii*) Relaying only,
(*iii*) Sensing relaying, and
(*iv*) Inactive.

In the sensing state, the node probes the environment and generates data at a constant rate. In the relaying state, the node does not sense the target but its communications circuitry is on to relay the data from other active nodes. On the other hand when a node is both sensing and relaying messages from other nodes, it is considered in the sensing-relaying state. Or else, the node is considered inactive and can turn off its sensing and communication circuitry.

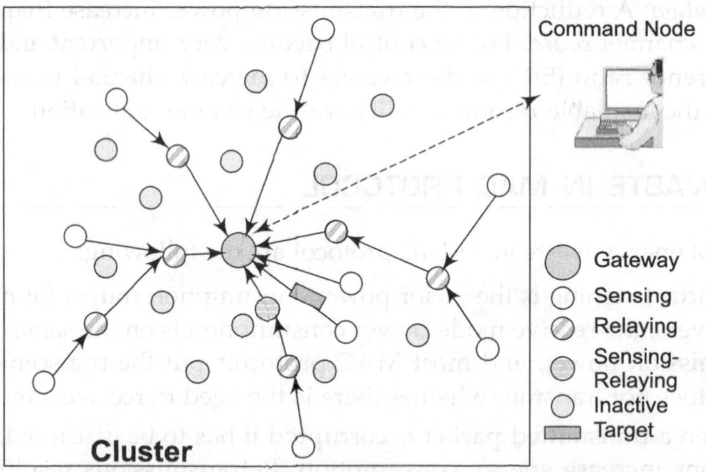

Fig. 7.4 A typical cluster in a sensor network

Fig. 7.4 shows an example of the state of sensors and routes within a typical cluster for a target-tracking application. A cost function is defined between any two nodes in terms of energy consumption, delay optimization and other performance metrics. This cost function is a least-cost path is found between sensor nodes and the gateway. The gateway will continuously monitor the available energy level at every sensor that is active in data processing, sensing, or in forwarding data packets, relaying. Based on application rerouting is triggered different set of sensors to probe the environment or the depletion of the battery of an active node.

7.8 ENERGY CONSUMPTION MODELS

The energy consumption model is explained as follows:

7.8.1 The Classical Energy Consumption Model

Heinzelman et. al proposed an energy consumption model for sensors based on the observation that the energy consumption would likely be dominated by the data communications subsystem. Table 7.1 reproduces their model.

Table 7.1 Radio Characteristics, Classical model

Radio Mode	Energy Consumption
Transmitter Electronics ($E_{Tx\text{-}else}$) Receiver Electronics ($E_{Rx\text{-}else}$) ($E_{Tx\text{-}else} = E_{Rx\text{-}else} = E_{else}$)	50nJ/bit
Transmit Amplifier (e_{amp})	100pJ/bit/m^2
Idle (E_{idle})	40nJ/bit
Sleep	0

The model considers a low power consumption radio that was slightly better than some standard definitions, like Bluetooth. When computing node energy consumption, the CPU and the sensors are consumers that and it depend on the nature of the application. The radio model mode is used with figure of the energy consumption for study.

7.8.2 μAMPS Specific Model

Shih et. al presented a model developed for a specific platform, the μAMPS Wireless Sensor Node. The platform has a StrongARM SA 1110 microprocessor with a clock speed from 59 Mhz to 206 Mhz. The model takes into consideration the energy consumed by the microcontroller, energy lost due to leakage and the average consumption of the radio. Table 7.2 summarizes the model characteristics.

Table 7.2 Sensor states for μAMPS Model

Stat	SA-	Sense	R	Pk
Activ	Active	Sense	t	1040
Rea	Idle	Sense	r	400
Moni	Sleep	Sense	r	270
Obs	Sleep	Sense	o	200
Dee	Sleep	off	o	10

The μAMPS model doesn't specify the power consumed in transmitting or receiving one bit. Nonetheless, the platform uses transmission rate of 1 Mbps, so one can calculate the energy required for transmitting one bit, following a method based in the approach presented by Hill et. al. The energy used in transmitting or receiving one bit and is found by using the power value.

$$\text{Energy} = \text{Power} * \text{Time} \qquad ...(2.1)$$

Where, Power is in Watts and Time is in seconds

The difference between μAMPS model and the classical model presented is on the order of magnitude for the transmission case and on the order of magnitude for the receiving case.

7.8.3 Mica2 Specific Model

Polastre et. al proposed a model that presents the total energy consumption for Mica2 as the summation of energy transmitting, receiving, listening, sampling data and sleeping. Values are calculated using the expected consumption of the CPU and the radio, which can be found in specific datasheets. Table 7.3 presents a summary of current consumption.

Table 7.3 Current consumption for Mica2 Model

Operat	T	
Initializ		
Turn		
Switch		
Time		
Evalua		
Receiv		
Trans		
Sampl		

The current consumption and time, and assuming that Mica2 is powered by a 3V source, one can calculate energy in transmitting and receiving one bit, as:

$$\text{Energy} = \text{Current} * \text{Voltage} * \text{Time} \quad \text{...(2.2)}$$

Where, current is in Amperes, Voltage is in Volts and Time is in seconds.

With the μAMPS model, energy for transmission is comparable, while energy for reception is one order of magnitude bigger in the Mica2 case.

7.8.4 Mica2 Specific Model with Actual Measurements

Shnayder et.al presented a current consumption model based on measurements on the Mica2 platform [26]. A summary of the model is shown in Table 7.4.

Table 7.4 Current Consumption with Actual Measurements

Mode	Current	Mode	Current
CPU		Radio	
Active	8.0 μA	Rx	7.0 μA
Idle	3.2 μA	Tx (–20 dBm)	3.7 μA
ADC Noise Reduce	1.0 μA	Tx (–19 dBm)	5.2 μA
Power-down	103 μA	Tx(–15 dBm)	5.4 μA
Power-save	110 μA	Tx(–dBm)	6.5 μA
Standby	216 μA	Tx(–dBm)	7.1 μA
Extended Standby	223 μA	Tx(+dBm)	8.5 μA

Internal Oscillator	0.93 μA	Tx(+dBm)	11.6 μA
LEDs	2.2 μA	Tx(+dBm)	13.8 μA
Sensor board	0.7 μA	Tx(+dBm)	17.4 μA
EEPROM access		Tx(+10 dBm)	21.5 μA
Read	6.2 μA		
Read Time	565 μs		
Write	18.4 μA		
Write Time	12.9 μs		

Values presented in the table are calculated independently. The total current is found by summing the consumption for each active component. As an example, in calculating energy per bit transmitted and received, one may include only the CPU in the active state and presume the radio is transmitting with a certain power.

SUMMARY

The energy wastage of sensors discussed, here. Sensor lifetime as mentioned in this chapter highlighted the software and other hardware components. The estimate sensor lifetime, is limited by energy storage capacity. The measurements of energy consumption rate taken for Mica2 motes Sensor lifetime are limited by software and other hardware components. In this platform, CPU energy consumption in active state is three orders of magnitude smaller than energy spent in transmitting or receiving a bit. Chapter also illustrates energy consumption statistics when different transmission power levels are used.

QUESTIONS

1. Why power is important issue in wireless sensor networks?
2. What are the characteristics of power management in WSN?
3. Explain the reasons for energy management in wireless sensor network.
4. How the clustering has an impact of energy management in sensor network?
5. Write a note on μAMPS.

BIBLIOGRAPHY

- R. Ramanathan and R. Rosales-Hain, "Topology Control of Multi-hop Wireless Networks using Transmit Power Adjustment," Proc. of IEEE **INFOCOM,** 2000, pp. 404-413.
- Wei Ye, John Heidemann and Deborah Estrin. An Energy-EfficientMAC Protocol for Wireless Sensor Networks, In Proceedings of the 21st International Annual Joint Conference of the IEEE Computer and Communications Societies (INFOCOM 2002), New York, NY, USA, June, 2002.
- Akyildiz, W. Su, Y. Sankarasubramaniam, and E. Cayirci, "A survey on sensor networks," IEEE Communication Magazine, vol. 40, no. 8, pp. 102-116, Aug. 2002.
- W. Ye, J. Heidemann, and D. Estrin, "An energy-efficient MAC protocol for wireless sensor networks," in Proceedings of the IEEE Infocom, USC/Information Sciences Institute. New York, NY, USA: IEEE, June 2002.
- Y.-C. Tseng, C.-S. Hsu, and T.-Y. Hsieh, "Power-saving protocols for ieee 802.11-based multi-hop ad hoc networks," in Proc. of INFOCOM, vol. 1, 2002, pp. 200-209.

- T. Šimunic, L. Benini, P. Glynn, and G. de Micheli. "Event-driven power management". Proceedings of International Symposium on System Synthesis, 1999.
- Carlos Villavieja Prados, Loreto Mateu, Isaac Gelado, Ferran Martorell, Francesc Moll, Nacho Navarro, Marisa Gil. Experimental Runtime Power Consumption Measurements in Wireless Sensor Networks. Ubiquitous Computing and Ambient Intelligence (UCAmI), Granada, Spain, September 2005.
- T. Facchinetti, L. Almeida, G. C. Buttazzo, and C. Marchini. Real-time resource reservation protocol for wireless mobile ad hoc networks.
- J. Gomez, A. T. Campbell, M. Naghshineh, and C. Bisdikian, "Conserving Transmission Power in Wireless Ad-Hoc Networks," Proc. of IEEE Conference on Network Protocols (ICNP'01), November 2001, pp. 11-14.
- S. Narayanaswamy, et. al., "Power Control in Ad-hoc Networks: Theory, Architecture, Algorithm and Implementation of the COMPOW Protocol," Proc. of European Wireless Conference, 2002, pp. 156-162.
- M. T. Schmitz and B. M. Al-Hashimi. "Considering power variations of DVS processing elements for energy minimisation in distributed systems". Proceedings of the International Symposium on Systems Synthesis, ISSS, pages 250-255, 2001.

8

Localization in WSN

8.1 INTRODUCTION

Wireless sensor networks (WSNs) are an exciting concept of a wireless network having a wide variety of promising applications in real life. Wireless sensor networks are particularly interesting in hazardous or remote environments, or when a large number of sensor nodes have to be deployed for sensing, establishing wireless communication between each other and doing computational.

The localization subject is important research area where, there is an uncertainty about some positioning. If the sensor network is used for monitoring the temperature in a building, it is likely to know the exact position of each node. On the contrary, if the sensor network is used for monitoring the temperature in a remote forest, nodes may be deployed from an airplane and the precise location of most sensors to be traced out as shown in Fig. 8.1.

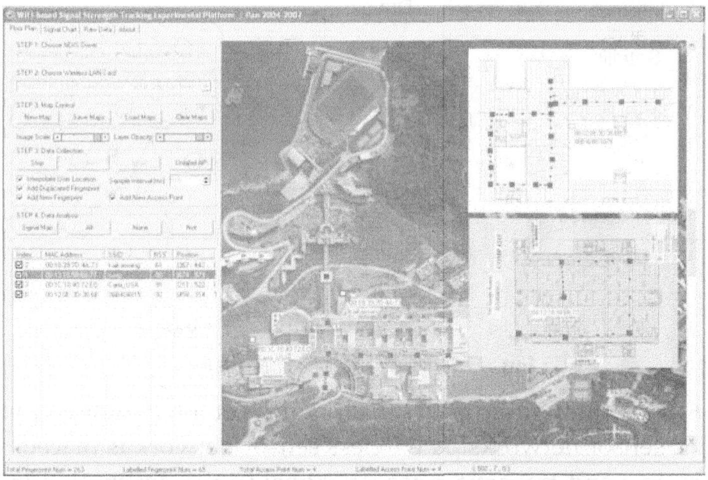

Fig. 8.1 Localization of global infrastructure

Here, sensor networks are being used in large number of military and civil needs such as surveillance of armed troops their vehicles in battlefields as well as for detection, tracking and classification of enemy targets. Today WSN is a key technology for different types of environments where large number of sensor nodes have to be placed in a given location.

One of the important technical issues that characterize the functionality of WSN networks is the self-localization of nodes. It is often necessary that the data transmitted to the sink of the network from source node are accompanied by the location information. Schemes for localization in WSN and numerous studies have been performed for a civil use. The localization system should be cheap, power-aware and variably accurate depending on the application. An effective localization algorithm can use all the available information from the motes to compute all the positions.

Presently, all the localization techniques suffer from one or the other problem related to accuracy, range, distribution and area. Other problems like energy efficiency and power consumption typically focus on minimizing the transmission energy for long range applications, where the transmission energy is a dominant factor in the total energy consumption spectrum. However in short range applications for wireless sensor networks, the total energy consumption is comparable and even dominates the transmission energy.

8.2 LOCALIZATION

WSN play a major role towards sensing and computing within human supervision. Location awareness is vital for wireless sensor networks, since many applications like environmental monitoring, vehicle tracking and mapping depend on knowing the exact location of sensor nodes. Localization is done when there is an uncertainty regarding location. The increasing miniaturization in the semiconductor field is lead to the evolution of very small and low-cost sensors. Their sizes are small and they are strongly limited with respect to processor capacity, memory size and energy resources.

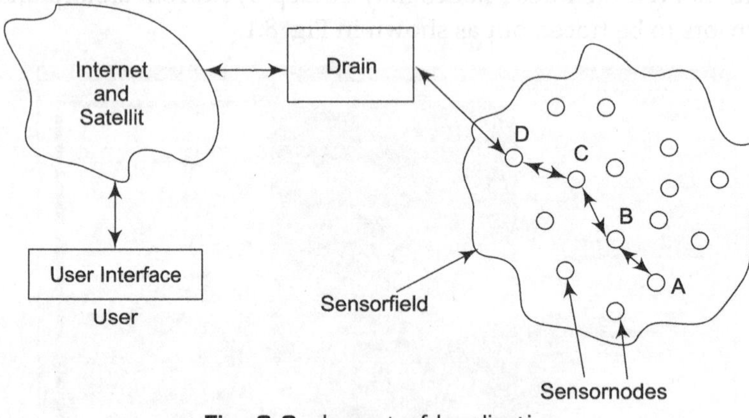

Fig. 8.2 Layout of localization

As shown in Fig. 8.2 several thousands of sensor nodes get into wireless contact with each other and form large ad-hoc sensor networks. A wireless sensor network is placed to monitor different environmental parameters and transmit a beacon signal to the destination in the laid infrastructure node.

Possible positioning technologies are the Global Positioning System (GPS) or the Global System for Mobile Communication (GSM) in WSN enable detection of wood fire or monitoring of artificial dikes applications. The resulting data are only meaningful when combined with the geographical position of the sensor.

Recently, many localization techniques have been proposed to allow the nodes to estimate their own locations using information transmitted to know their positions. The sensor network energy consumption includes the energy consumed by all the sensor nodes along with the signal path which also depends on distance and placement of nodes.

Localization refers to identifying which nodes have a priori known locations (called reference nodes) and which nodes do not (called unknown nodes). The unknown nodes may be cooperative or non-cooperative. Non-cooperative nodes cannot participate actively in the localization algorithm.

In most cases, the location information is needed for all unknown nodes at the very beginning of network operation. The location information of nodes in the network is fundamental for a number of reasons:

(*i*) To provide location stamps.
(*ii*) To locate and track point objects.
(*iii*) To determine the quality of coverage.
(*iv*) To achieve load balancing.
(*v*) To form clusters.
(*vi*) To facilitate routing.
(*vii*) To perform efficient spatial querying.
(*viii*) To monitor the spatial evolution of a diffuse phenomenon.

8.3 NEED OF LOCALIZATION

Localization in WSN is an active area of research and so there exist literature surveys on this topic. A localization algorithm is to find out the location of the sensor nodes relatively or absolutely. Here, pointed out some of the following uses of localization:

I. Localization enables the efficient routing: A sensor network has large number of nodes that communicate with other sensors placed at very short distance about a few meters. The data sensed by a node communicated to the central unit (or a sink) through several other nodes thru multi-hop routing to identify their relative position with respect to their neighbors.

II. Localization provides the power saving: In the application of deployed sensor scenario the nodes collect information and send to the source node as per the topology structure. Thus to save power neighboring nodes combines the data and then communicate the combined, that reduced data set, thereby conserving power with data fusion.

III. Localization assists in the applications like target tracking: In the application, where there is a need to determine the range, speed and the direction of the target. Sensors are deployed in order to calculate the global orientation of the target to know the location of the sensor nodes.

IV. Localization useful in locating the source of the data: In many applications, networks wireless sensor nodes are used. The nodes are normally in sleep mode and are awakened. The node that sense and transmit the data requires a location stamp and therefore localization can be determined.

V. Localization performance pertains to the resolution of location information desired. One such example as shown in Fig. 8.3 is the use of globally accessible beacons or expensive GPS techniques to localize individual sensor in the world wide accepted coordinate system and different environments. Here, the localization technique provides absolute(x, y, and z) coordinates, or perhaps it will suffice to provide relative coordinates or symbolic locations.

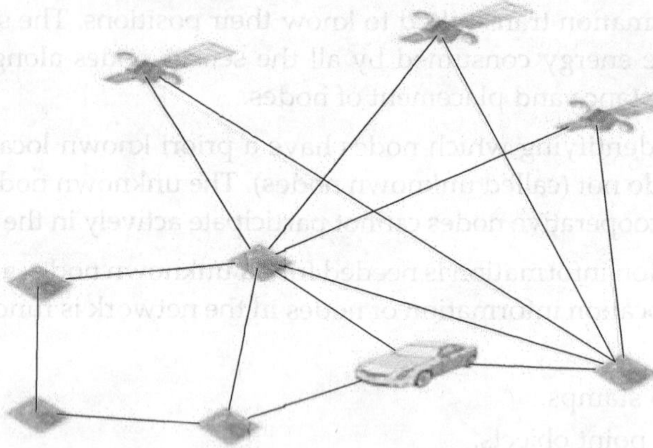

Fig. 8.3 GPS technique to localize sensor

The basic localization algorithm may be based on a number of techniques, such as proximity, calculation of centroid's, constraints, ranging, angulations, pattern recognition, multi-dimensional scaling, and potential methods.

8.4 DESIGN RESTRAINTS IN LOCALIZATION ALGORITHM

Localization algorithms vary depending on applications and its performance will depend on critical sensor network parameters, such as the radio range, the density of nodes, the anchor-to-node ratio, and the solution that gives adequate performance over a range of reasonable parameter values.

For many applications like habitat monitoring, smart buildings, failure detection and target tracking, it is necessary to accurately orient the nodes with respect to a globally recognized coordinate system in order to record the data. The goal is to be able to position nodes with a given accuracy, or to classify nodes as being" non-localizable".

The theory of localization justifies the concept as a mechanism to obtain location based information of a sensor in a coordinate system with respect to specific time, distance, received signal strength, time difference of arrival, angle of arrival, time of flight and lifetime related parametric values to locate the position of a particular object.

I. Resource constraints: Sensor networks are typically quite resource starved *i.e.,* it is battery power constraint. This means communication processing and sensing consume power and it actively reduces the lifespan of the node. As sensor are placed in a large scale along with hundreds or thousands of real working nodes in an environment. Hence the nodes must be cheap to fabricate with good power backup and easy to deploy.

II. Density of node: Node density is another important parameter to which localization algorithms are sensitive. Multiple hop count based schemes generally require high node density so that the hop count approximation distance becomes accurate. Thus while designing or analyzing an algorithm it is quite important to observe the throughput for high node density.

III. Non convex topologies: In localization algorithms border nodes are a cause of concern as proposed by Chen et al., 2006. It is because less information is available through them and the information received is also of a lower quality. The problem can be sorted, when a sensor network has a non convex shape which is not always desirable. Hence, shape of a sensor network plays a major role in collection of useful data.

IV. Terrain irregularities and environmental design: There are many environmental effects and irregularities that greatly affect the localization. Large buildings can block the line of sight or interfere with signals and introduce errors. So, the real deployment of sensors in the sensor network plays the significant role.

V. System architecture: Design of system architecture is the most important issue, where the topology control should be as per the algorithm selection. Centralized or Distributed algorithms are applied as per the topology layout in the applications.

8.5 LOCALIZATION METHODS

There are many localization algorithms for sensor networks have been proposed to provide per-node location information. Generally speaking, there are two approaches to localization are:-

I. *Range free method or Coarse-grained localization using minimal information:* These typically use a small set of discrete measurements, such as the information used to compute location

II. *Range based method or Fine-grained localization using detailed information:* These are typically based on measurements, such as RF power, signal waveform, time stamps, etc. that is either real-valued or discrete with a large number of quantization levels. These include techniques based on radio signal strengths, timing information, and angulations.

The tradeoff that emerges between the two approaches is easy to see: while former has minimal information techniques are simpler to implement, and likely involve lower resource consumption and later is having equipment costs, they provide lower accuracy than the detailed information techniques.

8.5.1 Range Free Method or Coarse-Grained Localization Using Minimal Information

This is also known as Range-Free Localization Schemes. The range free localization technique does not use the channel model for calculating the location. It makes no assumption about the availability or validity of such information. WSN devices are having hardware limitations so; range-free localization is being pursued as a cost-effective alternative to more expensive range-based approaches. The range free localization technique does not use the channel model for calculating the location.

The Range-Free Localization protocols algorithms are implemented in accordance with the design and it include:

I. Binary Proximity
II. Centroid Localization
III. Geometric Constraints
IV. Approximate Point In Triangle (APIT)
V. DV-Hop Localization
VI. Identifying Codes
VII. Amorphous Localization

I. Binary Proximity: Perhaps the most basic location technique is that of binary proximity - involving a simple decision of whether two nodes are within reception range of each other. A set of references nodes in the environment is non-overlapping mode. Either the reference nodes periodically emit beacons, or the unknown node transmits a beacon when it needs to be localized. If reference nodes emit beacons, these include their location IDs. The unknown node must then determine which node it is closest to, and this provides coarse grained localization. Technique can be of considerable use in practice.

II. Centroid Localization: This range-free localization protocol was proposed by N. Bulusu and J. Heidemann. The same proximity information can be used to greater advantage when the density of reference nodes is sufficiently high that there are several reference nodes within the range of the unknown node. This simple centroid technique has been investigated using a model with each node having a simple circular range R in an infinite square mesh of reference nodes spaced a distance d apart.

In this proximity-based, coarse grained localization algorithm, that uses anchor beacons, containing location information (X_i, Y_i), to estimate node position. After receiving these beacons, a node estimates its location using the following centroid formula:

$$(X_{en}, Y_{en}) = \left(\frac{X_1 + \ldots + X_N}{N}, \frac{Y_1 + \ldots + Y_N}{N} \right)$$

The distinguished advantage of this Centroid localization scheme is its simplicity and ease of implementation.

III. Geometric Constraints: Fig. 8.3 illustrates the use of intersecting geometric constraints for localization. If the bounds on radio or other signal coverage for a given node can be described by a geometric shape, this can be used to provide location estimates by determining which geometric regions that node is constrained to be in, because of intersections between overlapping coverage regions. Although, arbitrary shapes can be potentially computed in this manner, a computational simplification that can be used to determine this bounded region is to use rectangular bounding boxes as location estimates.

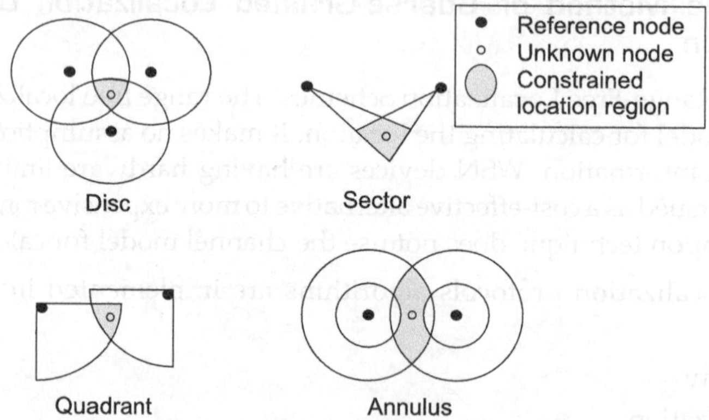

Fig. 8.4 Localization using intersection of geometric constraints

Localization techniques using such geometric regions were first described by Doherty et al. One of the nice features of these techniques is that not only the unknown nodes can use the centroid of the overlapping region as a specific location estimate if necessary, but they can also determine a bound on the location error using the size of this region. When the upper bounds on these regions are tight, the accuracy of this geometric approach can be further enhanced by incorporating "negative information" about which reference nodes are not within range.

IV. Approximate Point In Triangle (APIT): A related approach to localization using geometric constraints is the approximate point-in-triangle (APIT) technique is illustrated in Fig. 8.4. APIT is similar to the above techniques in that it provides location estimates as the centroid of an intersection of regions. Its novelty lies in how the regions are defined - as triangles between different sets of three reference nodes (rather than the coverage of a single node).

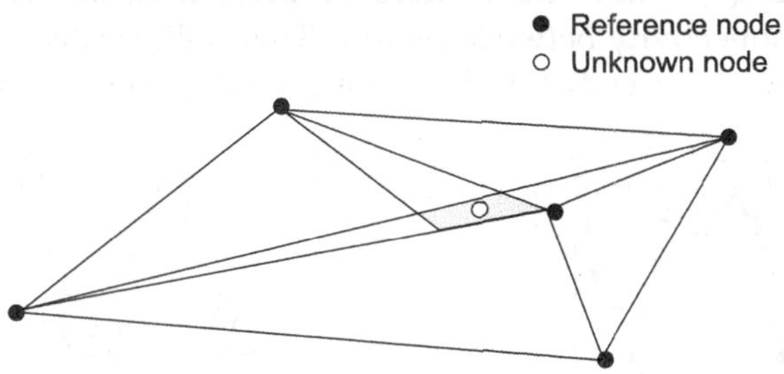

Fig. 8.5 The approximate point-in-range (APIT) technique

V. DV-Hop localization: DV-Hop localization is proposed by D. Niculescu and B. Nath in the Navigate project. DV-Hop localization uses a mechanism that is similar to classical distance vector routing. In this work, one anchor broadcasts a beacon to be flooded throughout the network containing the anchors location with a hop-count parameter initialized to one. Each receiving node maintains the minimum counter value per anchor of all beacons it receives and ignores those beacons with higher hop-count values. The beacons are flooded outward with hop-count a value incremented at every intermediate hop is shown in Fig. 8.6. The hop count for a single anchor A, generated by simulation this mechanism, all nodes in the network get the shortest distance, in hops, to every anchor.

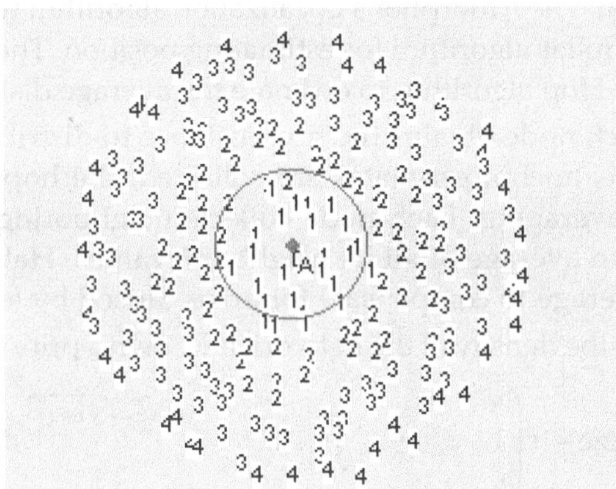

Fig. 8.6 Anchor beacon propagation phase

The average single hop distance is then estimated by anchor i using the following formula where, In this formula, (xj,yj) is the location of anchor j, and hj is the distance, in hops, from anchor j to anchor i. Once calculated, anchors propagate the estimated HopSize information out to the nearby nodes. Theoretically, if errors exist in the distance estimation, the more anchors a node can hear the more precise localization.

$$\text{HopSize}_i = \frac{\sum \sqrt{\left(x_i - x_j\right)^2 + \left(y_i - y_j\right)^2}}{\sum h_j}$$

VI. Identifying Codes: In this technique, referred to as the identifying code construction (ID-CODE) algorithm, the sensor deployment is planned in such a way as to ensure that each resolvable location is covered by a unique set of sensors. The algorithm runs on a deployment region graph G= (V, E) in which vertices V represent the different regions, and the edges E represent radio connectivity between regions This is illustrated in Fig. 8.7. The algorithm ID-CODE is a polynomial greedy heuristic that provides good solutions in practice.

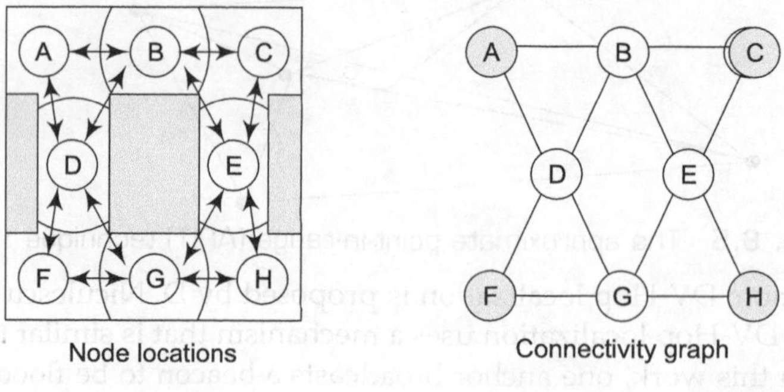

V:	A	B	C	D	E	F	G	H
ID:	A	A, C	C	A, F	C, H	F	F, H	H

Fig. 8.7 Illustration of the ID-CODE technique showing uniquely identifiable regions

VII. Amorphous localization: The Amorphous Localization algorithm was proposed independently from DV-Hop uses a similar algorithm for estimating position. The algorithm takes a different approach from the DV-Hop algorithm to estimate the average distance of a single hop.

First, like DV-Hop, each node obtains the hop distance to distributed anchors through beacon propagation. Once, anchor estimates are collected, the hop distance estimation is obtained through local averaging. Each node collects neighboring nodes' hop distance estimates and computes an average of all its neighbors' values. Half of the radio range is then deducted from this average to compensate for error caused by low resolution.

The work assumes that the density of the network is known a priori, so the HopSize is given by:

$$\text{HopSize} = r\left(1 + e^{-n_{local}} - \int_{-1}^{1} e^{-\frac{n_{local}}{\pi}\left(\arccos t - t\sqrt{1-t^2}\right)} dt\right)$$

Finally, after obtaining the estimated distances to three anchors, triangulation is used to estimate a node's location.

8.5.2 Range Based Method or Fine-Grained Localization Using Detailed Information

This is also known as Range based Localization Schemes. The range based model protocol use absolute point-to-point distance estimates (range) or angle estimates for calculating location. It is basically uses the channel characteristics to find the distance. The range method exploits information about the distance to neighboring nodes. The distances cannot be measured directly so theoretically, use the following methods:

I. Received signal strength indication (RSSI),

II. Time of arrival (TOA), and

III. Angle of arrival (AOA).

These include triangulation using distance estimates, pattern matching, and sequence decoding. Although used in the large-scale GPS, signals are not capable of providing precise distance estimates over short ranges typical of WSN, because of synchronization limitations.

I. The Received signal strength indication (RSSI): The RSSI method able to find the relative distances and location of the other nodes with the use one master node that knows its position. The RSSI method is widely used; and it exploits beacons and using the received power from the beacons to finds the distance. RSSI suggests an elegant solution to the hardware ranging problem, all sensor nodes are likely to have radios signal information. This technique uses the three different operations to locate all the sensors.

The major steps in RSSI algorithm are:

(*i*) Single sensor identification.

(*ii*) Imprecise sensor location map definition.

(*iii*) Sensor location map refinement.

First a beacon initiates the localization algorithm and transmits the signal. Nearby sensors will receive this signal and then they send an acknowledgment. The master then gives a unique ID to each of these nodes. The nodes calculate the distance based on receive power and transmit this information to the master. This procedure continues until the complete network is covered.

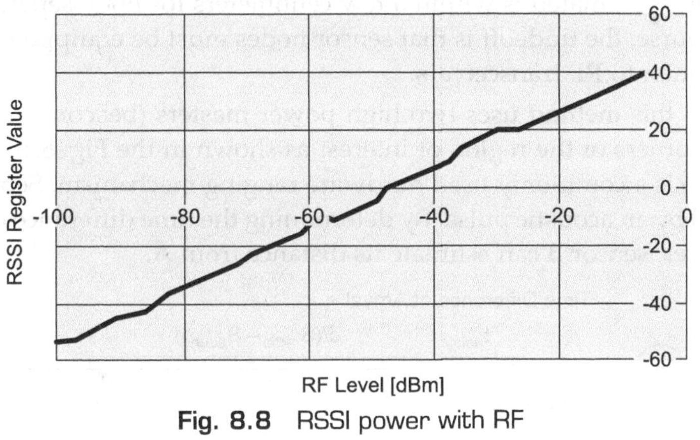

Fig. 8.8 RSSI power with RF

RSSI is a relatively cheap solution without any extra devices, as all sensor nodes are likely to have radios. It requires high power beacons and a lot of power is wasted in transmitting the relative distances to the master as shown in Fig. 8.8. The relative distances level and the RSSI register value show the gradual rise.

To a first-order approximation, mean radio signal strengths diminish with distance according to a power law. RSS-based ranging may perform much better in situations, where the fading effects can be combated by diversity techniques that take advantage of separate spatio-temporally correlated signal samples

II. Time of arrival (TOA): A more promising technique is the combined use of ultrasound/acoustic and radio signals to estimate distances by determining the TDOA of these signals. These methods record the time-of-arrival (ToA) or time-difference-of-arrival (TDoA), where the propagation time can be directly translated into distance, and signal propagation speed. This technique is conceptually quite simple, and is illustrated in Fig. 8.9. The idea is to simultaneously transmit both the radio and acoustic signals (audible or ultrasound) and measure the times Tr and Ts of the arrival of these signals respectively at the receiver.

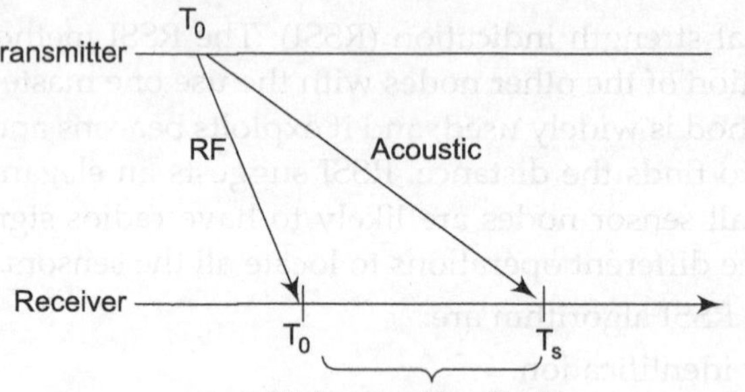

Fig. 8.9 Time-difference-of-arrival (TDoA)

Since, the speed of the radio signal is much larger than the speed of the acoustic signal, the distance is then simply estimated as $(T_s - T_r) * V_s$, where V_s is the speed of the acoustic signal. One minor limitation of acoustic ranging is that it generally requires the nodes to be in fairly close proximity to each other (within a few meters) and preferably in line of sight. On the whole, acoustic TDOA ranging techniques can be very accurate in practical settings. For instance, distance estimated is within a few centimeters for node separations to fewer than 3 meters. Of course, the tradeoff is that sensor nodes must be equipped with acoustic transceivers in addition to RF transceivers.

For example in this method uses two high power masters (beacons), and two masters are placed in the corners of the region of interest as shown in the Fig. 8.10. Time Difference of Arrival (TDoA) is a commonly used hardware ranging mechanism. Sensor A sends a radio pulse followed by an acoustic pulse. By determining the time difference between the arrivals of the two pulses, sensor B can estimate its distance from A.

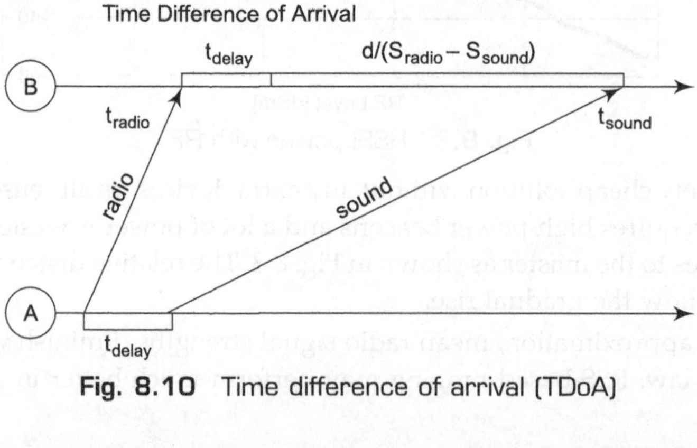

Fig. 8.10 Time difference of arrival (TDoA)

The unknown node detects the signal strength from the two masters and immediately calculates its location based on circular triangulation. Finally, when the acknowledgment from all the nodes reaches the beacon, the localization of the node is evaluated.

This method requires less power than the one master method. TDoA methods are impressively accurate under line-of-sight conditions. But this line-of-sight condition is difficult to meet in some environments.

The disadvantage is that it requires high power beacons. Each beacon should have enough power to reach the farthest node. Moreover, to solve the unique location of two beacon nodes by circular triangulation is by restricting the beacon node to be at the corners of the boundary.

TDoA methods are impressively accurate under line-of-sight conditions and it is more effective in practice than RSS. It is due to the difference between using signal travel time and signal magnitude, where the former is vulnerable only to difference while the latter is vulnerable to both difference and multipath.

III. Angle of Arrival (AOA): Another possibility for localization is the use of angular estimates instead of distance estimates. Angles can potentially be estimated by using rotating directional beacons, or by using nodes equipped with a phased array of RF or ultrasonic receivers. AoA is a technique that estimates the angle at which signals are received and calculate node positions with the help of simple geometric relationships. Generally, AoA techniques provide more accurate localization result than RSSI based techniques but the cost of hardware of very high in AoA.

This method is using two high power beacons, where the nodes are mobile. Localization result in power wastage by the beacon as well as nodes. In these methods, several spatially separated microphones hear a single transmitted signal. And by analyzing the phase or time difference between the signal's arrivals at different microphones, it is possible to discover the angle of arrival of the signal.

However, angle-of-arrival hardware tends to be bulkier and more expensive than TDoA ranging hardware, since each node must have one speaker and several microphones. Furthermore, the need for spatial separation between speakers is difficult to accommodate as the form factor of sensors shrinks.

Angle of Arrival hardware is sometimes augmented with digital compasses. A digital compass simply indicates the global orientation of its node, which can be quite useful in conjunction with AoA information.

8.5.3 Pattern Matching (RADAR)

An alternative to measuring distances or angles that is possible in some contexts is to use a predetermined "map" of signal coverage in different locations of the environment, and use this map to determine, where a particular node is located by performing pattern matching on its measurements. An example of this technique is RADAR.

8.5.4 RF Sequence Decoding (Ecolocation)

The ecolocation technique uses the relative ordering of received radio signal strengths for different references as the basis for localization.

8.6 LOCALIZATION ALGORITHM

There are several forms of localization algorithm research study carried out in different scenarios. One such algorithm for the wireless nodes are mobile is mentioned below. The area for the wireless mobile sensor network is less very when compared with other networks like fixed node networks and wired networks This localization algorithm defined to cover a particular geographic location with sensors covering a span of maximum area.

Step 1 – defines nodes: M, N_1 and N_2; Define range: R

Step 2 – compare distance among nodes

Step 3 – if $N_1 < R$; place $N_1 = R$ from M

Step 4 – compare distance between M and N_2 If $N_2 > R$; then compare distance between N_1 and N_2

Step 5 – if $N_2 < R$; place N_2 from N_1 at range R

In the one main node which is connected to the system and two sub nodes. All are randomly placed in a particular area. As shown in Fig. 8.11 there are main node and pair of sub nodes. Firstly, we define node M which is the main node connected to system. N_1 and N_2 are randomly placed sub nodes. The sub nodes represent sensors and have sensing some range.

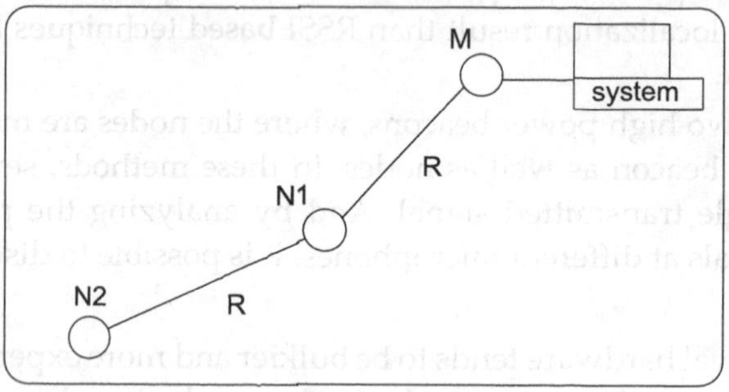

Fig. 8.11 Placement of sensor nodes after applying algorithm for coverage of maximum area

In step one, distance between M and N_1 is compared, if it is less than the defined range, N_1 is placed at a distance R from M. In second step distance between M and N_2 is compared if it is greater than the defined range, then the distance between N_1 and N_2 is compared, if this distance is less than the defined range R, we have to place the sub node at a distance R from N_1. Using this approach localization of the sensors node is carried out.

The advantage of this algorithm is vital since, it requires lesser number of sensor nodes and still it can cover a larger area. This procedure is also cost effective as minimum number of sensors is used.

8.7 LOCALIZATION CATEGORY

There are different areas of localization in wireless sensor networks and can be classified into two broad categories.

I. *Centralized Localization*: Centralized localization is basically migration of inter-node ranging and connectivity data to a sufficiently powerful central base station and then the migration of resulting locations back to respective nodes. The advantage of centralized algorithms are that it eliminates the problem of computation in each node, at the same time the limitations lie in the communication cost of moving data back to the base station.

II. *Distributed Localization*: In Distributed localizations all the relevant computations are done on the sensor nodes themselves and the nodes communicate with each other to get their positions in a network. Distributed localizations can be categorized into three classes.

(*a*) Beacon-based distributed algorithms: Beacon-based distributed algorithms start with some group of beacons and nodes in the network to obtain a distance measurement to a few beacons, and then use these measurements to determine their own location.

(*b*) Relaxation-based distributed algorithms: In relaxation-based distributed algorithms use a coarse algorithm to roughly localize nodes in the network. This coarse algorithm is followed by a refinement step, which typically involves each node adjusting its position to approximate the optimal solution.

(*c*) Coordinate system stitching based distributed algorithms: In Coordinate system stitching the network is divided into small overlapping sub regions, each of which creates an optimal local map. Next the scheme merges the local maps into a single global map.

8.8 TAXONOMY OF LOCALIZATION SYSTEMS

Taxonomy of localization system, categories is partitioned into "active" and "passive" mode.

8.8.1 Active Localization

Active localization techniques emit signals into the environment that are used to measure range to the target. These signals may be emitted by infrastructure components or by targets. Active localization is subdivided into three subcategories:

8.8.1.1 Non-cooperative

In an active, non-cooperative system, system elements emit ranging signals that are distorted or reflected by passive elements. The system elements receive the signals and analyze them to deduce their location relative to passive elements of the environment. Examples in radar systems and reflective sonar systems often used in robotics.

8.8.1.2 Cooperative Target

In a cooperative target system, the targets emit a signal with known characteristics, and other elements of the system detect the signals and use information about the signal arrivals to deduce the target's location. The cooperative target system also involves some synchronization mechanism to readily compute signal time of arrival (ToF).

8.8.1.3 Cooperative Infrastructure

In a cooperative infrastructure system, elements of the infrastructure emit signals that targets can receive. The infrastructure system structure is that its receivers can compute their own location passively, without requiring any interaction with the infrastructure.

8.8.2 Passive Localization

Passive localization techniques differ from active ones in that they discover ranges and locations by passively monitoring existing signals in a particular channel. The term "passive" emit the signals outside the channel that is primarily analyzed for time-of-flight measurement. It measures range by TDoA of ambient acoustic signals would still be considered passive.

8.8.2.1 Blind Source Localization

In a blind source localization system, a signal source is localized without any a priori knowledge of the type of signal emitted. Typically, this is done by "blind beam-forming", that effectively cross-correlates the signals from different receivers. These techniques generally only work so long as the signals being compared are "coherent", which in practice often limits the spacing of receivers because of signal distortion induced by the environment.

8.8.2.2 Passive Target Localization

Similar to blind localization, a passive target localization system is usually based on coherent combination of signals, with the added assumption of some knowledge of the source. By assuming a model for the signals generated by the source, filtering can be applied to improve the performance of the algorithms and to reduce the computational and communications requirements.

8.8.2.3 Passive Self-localization

In passive self-localization, existing beacon signals from known infrastructure elements are used by a target to passively deduce its own location.

In an ad-hoc setup, there is no guarantee that all the sensor nodes will not be in communication and sensing range to each other, nor that the sensing and communications properties will remain constant over time.

SUMMARY

In this chapter discussed the importance of localization and how such localization can be evaluated. The sensor devices envisioned and the estimation accuracy desired by location dependent applications, range-free localization schemes are regarded as cost-effective and sufficient solution form localization in sensor networks. Here, proposed system configurations of different recently proposed range-free localization schemes as a design guideline for further research.

Moreover, it provides insight of chapter focused on how localization error affects a variety of location-dependent applications. From a theoretical viewpoint, the fundamental characterization of error behavior under different measurement error distributions is still a topic of active research. Since, each application is likely to have its own requirements in terms of accuracy, latency and power consumption, node localization needs to be explored further in the context of each target application. The chapter also includes how to determining the geographic location of nodes in a sensor network is essential for many aspects of system operation: data stamping, tracking, signal processing, querying, topology control, clustering, and routing.

Finally, the chapter shows node localization is an application specific problem for which a one size fits all solution is unlikely to exist for all applications.

QUESTIONS

1. Explain the layout of the localization.
2. What is the need of localization?
3. What are the design restraints in localization algorithm?
4. Illustrates the localization methods.
5. Explain the localization algorithm.
6. Brief out the taxonomy of localization systems.

BIBLIOGRAPHY

- Biswas P and Ye Y (2006) A distributed method for solving semi definite programs arising from ad-hoc wireless sensor network localization, in Multiscale Optimization Methods and applications J. Nonconvex Optima. Appl. Vol. 82, pp. 69-84.
- A new approach for self localization of wireless sensor network O.P. Sahu and Tarun Dubey, Indian Journal of Science and Technology Vol.2 No. 11 (Nov. 2009).
- Wan Z, Zhang J and Zhu H (2008) on energy-efficient and low latency media access control protocol for wireless sensor networks. Proc. IEEE WCNC. Vol.5, pp. 148-164. Stephen Fitzpatrick and Lambert Meertens. Diffusion based localization. Private communication, 2004.
- C. Savarese, J. Rabaey, and J. Beutel. Locationing in distributed ad-hoc wireless sensor networks, 2001.
- L. Girod and D. Estrin, Robust Range Estimation using Acoustic and Multimodal Sensing, In Proceedings of IROS '01, Maui, Hawaii, October 2001.
- T. He, C. Huang, B. M. Blum, J. A. Stankovic and T. F. Abdelzaher, Range-Free Localization Schemes in Large Scale Sensor Networks, In Proceedings of MobiCom 2003.
- Localization Algorithms in Wireless Sensor Networks: Current Approaches and Future Challenges, Amitangshu Pal, Network Protocols and Algorithms, 2010, Vol. 2, No. 1.
- King-Yip Cheng, King-Shan Lui and Vincent Tam, "Localization in Sensor Networks with Limited Number of Anchors and Clustered Placement", in Proceedings of Wireless Communications and Networking Conference, 2007 (IEEE WCNC 2007), March 2007, pp. 4425 -4429.
- Real-time Localization for Wireless Sensor Networks with multiple beacon transmissions, G. S. Paschos, E. D. Vagenas, S. A. Kotsopoulos, Wireless Laboratory, University of Patras, Greece.
- Mondinelli F. and Kovacs-Vajna Z., "Self-Localizing Sensor Network Architectures", IEEE Transactions on Instrumentation and Measurement, Vol. 53, No. 2, April 2004.
- Sichitiu M. and Ramadurai V., "Localization of Wireless Sensor Networks with a Mobile Beacon", Mobile Ad-hoc and Sensor Systems Conference, October 2004.
- Wireless Sensor Networks: A Survey on Ultra-Low Power-Aware Design, Itziar Marín, Eduardo Arceredillo, Aitzol Zuloaga, and Jagoba Arias, World Academy of Science, Engineering and Technology, 2005.

WSN STANDARDS

9.1 INTRODUCTION

One of the biggest challenges facing companies deploying wireless sensors is the disparate standards, protocols and methods of communication and data formats. A standards body is typically slow and is to utilize as interoperability of standards to drive the adoption of wireless sensors.

The Table 9.1, short list some of the standards that is being used in wireless sensor networking. Standards are very important not only for the adoption of wireless sensor technology but also it leads to an eventual explosion of end products build on that standard which is a necessity for any emerging market.

Table 9.1 List of the wireless standards

Wireless Standards	Wireless Standards Description
802.11	A family of specifications developed by IEEE for local area networks. Typically, high bandwidth and high speed data rates and larger data packets.
802.15	A family of specifications developed by IEEE for personal-area networks. Typically low power, low rates, and small data packet size.
Bluetooth	A short range wireless standard operating in the unlicensed 2.4 GHz spectrum
ISA SP100	An open industrial wireless standard trying to support multiple protocols in a single standard. Includes 2.4 GHz and 802.14.5 radios.
Wi-Fi	Wireless Fidelity – technologies based on the 802.11 standard.
WiMax	World Interoperability for Microwave Access – a standard for wireless broadband over long distances based on IEEE 802.16
WirelessHART	An open wireless communication standard from the HART Communications Foundation designed for process measurement and control applications. Based on the 802.15 standard and frequency hopping spread spectrum technology
ZigBee	A low data rate, two way standard for home automation and data networks. Uses very low power consumption to create mesh networks using on 802.14 radios.

Zwave	Low power, low bandwidth communications standard designed for interoperability between systems and devices. Geared toward the residential and light commercial devices.

Recent advancements in information and wireless communication technologies are paving the way for new paradigms in embedded computing systems, and the integration of a wireless module make objects smarter. The network enable a wide range of new applications in the areas such as building automation, security, consumer electronics, industrial automation, process control, environmental control, etc. Wireless network standards such as ZigBee, Bluetooth, WirelessHART, etc. are all wireless standards and products more appropriate for industrial applications.

9.2 IEEE 802 STANDARDS

Wireless Sensor Networks design is influenced by several factors depending on the application of sensor networks and its network functionalities. There is a wide range of wireless communication protocol standards for WSN applications and their requirements in terms of low bandwidth, low energy consumption so that network/nodes lifetime is prolonged as much as possible. In fact, meeting energy requirements is most often the main goal of WSNs protocols and technologies.

The IEEE 802 main standards are shown in Fig. 9.1.

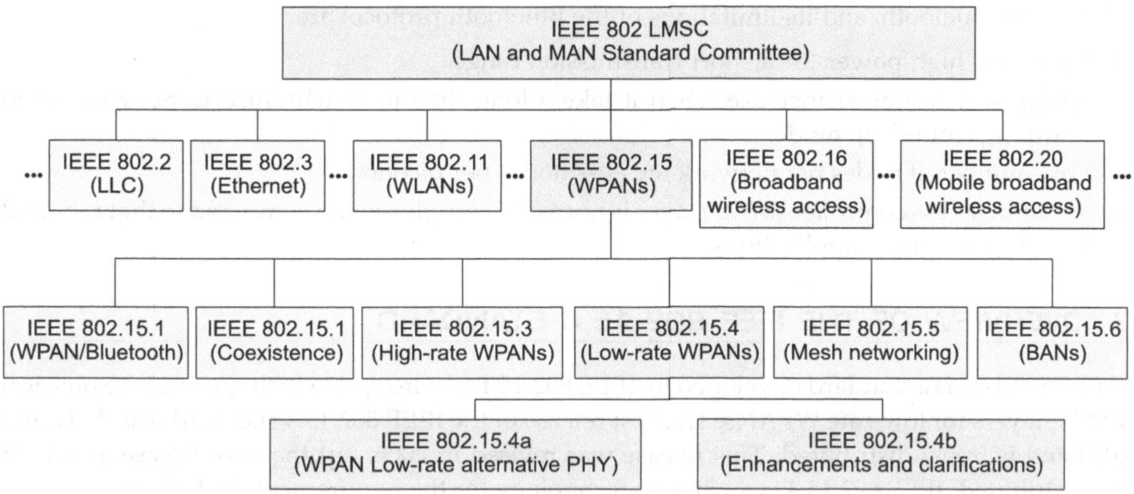

Fig. 9.1 IEEE 802 main standards

Wireless Sensor Networks are often associated to Personal Area Networks (PANs), networks for interconnecting devices centered on an individual person's workspace, typically in a short range. They can be both wired or wireless and they are called WPANs. Wireless Sensor Networks, due to their special characteristics, are actually belonging to the Low-Rate WPANs, for which IEEE 802.15.4 and ZigBee standards are two of the main standards.

The standard protocols, IEEE 802.15.4 and ZigBee, combined with commercial technologies as a baseline to enable WSN infrastructures capable of supporting the Quality of Service (QoS) requirements. The joint efforts of the IEEE 802.15.4 Task Group and the ZigBee Alliance have ended up with the specification of a standard protocol stack for Low-Rate Wireless Personal Area Networks (LR-WPANs), an enabling technology for Wireless Sensor Networks. Therefore, the IEEE 802.15.4 and

ZigBee protocols as a baseline to easier, faster and widespread development, deployment and adoption.

9.2.1 IEEE802.11x

IEEE802.11 is a standard for local area networking for relatively high bandwidth data transfer between computers or other devices. The data transfer rate ranges from as low as 1 Mbps to over 50 Mbps.

Typical transmission range is 300 feet with a standard antenna; the range can be greatly improved with use of a directional high gain antenna. Both frequency hopping and direct sequence spread spectrum modulation schemes are available. While the data rates are certainly high enough for wireless sensor applications, the power requirements generally preclude its use in wireless sensor applications.

9.2.2 Bluetooth (IEEE802.15.1 and .2)

Bluetooth is a personal area network (PAN) standard that is lower power than 802.11. Bluetooth uses a star network topology that supports up to seven remote nodes communicating with a single basestation. It originally speciûed for data transfer from personal computers to peripheral devices such as cell phones or personal digital assistants. While some companies have built wireless sensors based on Bluetooth, and its limitations of the Bluetooth protocol are:

I. Relatively high power for a short transmission range.
II. Nodes system power increase when it take a long time to synchronize to network when returning from sleep mode.
III. Low number of nodes per network are (<=7 nodes per piconet).
IV. Medium access controller (MAC) layer is normally complex when compared to that required for wireless sensor applications.

9.3 OVERVIEW OF THE IEEE 802.15.4 STANDARD

The IEEE 802.15.4 is a standard developed by IEEE 802.15 Task Group 4, which specifies the physical and MAC layers for low-rate WPANs. The first release of the IEEE 802.15.4 standard was delivered in 2003 and is freely distributed. This release was revised in 2006, but the new release is not yet freely distributed. IEEE 802.15.4 is a wireless technology for the requirements in WPAN.

IEEE 802.15.4 is a simple packet data protocol designed for lightweight wireless networks. Its architecture representation is shown in Fig. 9.2. This standard was not developed specially for Wireless Sensor Networks (WSN), but still can be used with WSNs, because the main requirements are related. Low power consumption, low cost, low data rate are typical requirements for WSNs.

The IEEE 802.15.4 protocol describes physical and Medium Access Control (MAC) layers. It is closely related with the ZigBee standard. The ZigBee Specification was released in 2004, which is publicly available and describes upper network and application layers.

Wireless channel access is controlled by MAC via Carrier Sense Multiple Access with Collision Avoidance (CSMA/CA) and optional time slotting. Communication between sensor nodes is performed with message Acknowledgment (ACK) and an optional beacon structure. Beacon-enabled and non-beacon operational modes are possible with the use of slotted and unslotted CSMA/CA accordingly. Multi-level security with 32-bit or 64-bit encryption can ensure security of data communication.

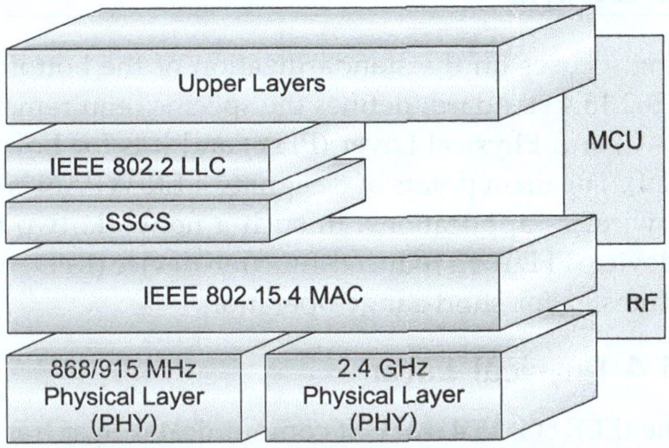

Fig. 9.2 IEEE 802.15.4 architecture

The IEEE 802.15.4 standard's main advantages are long battery life, selectable latency for controllers, sensors, remote monitoring and portable electronics. Configured for maximum battery life, it has the potential to last as long as the shelf life of most batteries. This is very important if a large number of node devices is used, where a frequent changing and recharging of batteries is impractical. Depending on the power consumption allowance, a transmission range can reach from 30 up to 100 meters and even more.

The architecture layers reach and include Logical Link Control (LLC). LLC is standardized in 802.2. LLC connects the MAC layer with upper layers through SSCS. Sometimes LLC is not used as separate layer, only some functions of LLC are implemented in software and are arbitrated by upper layers.

IEEE 802.15.4 supports multiple network topologies including Star, Cluster Tree and Mesh types. The PHY layer contains an RF Transceiver, operating with one of the specified frequency bands. The MAC layer provides access to the physical channel. Two modes of MAC operation are predefined in this standard: beacon-enabled and non- beacon enabled mode. Normally, PHY and MAC are parts of a transceiver, only control and data interchange are in responsibility of software in the MCU.

The standard supports the following characteristics:

1. Transmission frequencies, 868 MHz/902–928 MHz/2.48–2.5 GHz. it operates in one of the following three license - free bands:
   I. 868 – 868.6 MHz (e.g., Europe) with a data rate of 20 kbps.
   II. 902 – 928 MHz (e.g., North America) with a data rate of 40 kbps.
   III. 2400 – 2483.5 MHz (worldwide) with a data rate of 250 kbps.
2. Data rates of 20 Kbps (868 MHz Band) 40 Kbps (902 MHz band) and 250 Kbps (2.4 GHz band).
   I. Two addressing modes: 16 - bit short and 64 - bit IEEE addressing.
   II. Power management to ensure low - power consumption.
3. Supports star and peer-to-peer (mesh) network connections.
4. Standard specifies optional use of AES-128 security for encryption of transmitted data.

5. Link quality indication, which is useful for multi-hop mesh networking algorithms.
6. Uses direct sequence spread spectrum (DSSS) for robust data communications.

9.4 IEEE 802.15.4 LAYERS

The IEEE 802.15.4 Working focuses on the standardization of the bottom two layers of ISO/OSI protocol stack. The IEEE 802.15.4 standard, defines the specific requirements for the Wireless Medium Access Control (MAC) and Physical Layer (PHY) and it is for Low-Rate Wireless Personal Area Networks (LR-WPAN). The main points of designing a LR-WPAN is for the need of low-rate, low-power and low-cost wireless applications. In such a network, two types of devices can be involved: a full-function device (FFD) or a reduced-function device (RFD) which is easy to install, as it provides reliable data transfer for short-range operation.

9.4.1 IEEE 802.15.4 Physical Layer

PHY is the first layer in the IEEE 802.15.4 stack. It communicates over transmission media using 3 bands, which are divided into 27 channels. One band has a worldwide allowed carrier frequency of 2.4 GHz, 16 channels, 250 kbps bitrate with O-QPSK signal modulation.

The physical layer has the special purpose to perform channel analysis and to transmit/receive packets. It activates and deactivates the radio transceiver. A wireless sensor node can enter a sleep mode if the beacon-enabled mode is active. An example of the 2.4 GHz frequency band, divided into 16 channels is shown in Fig. 9.4.

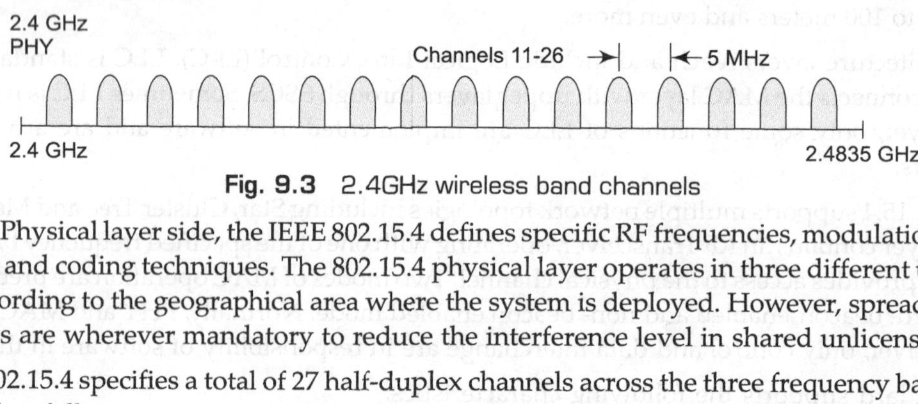

Fig. 9.3 2.4GHz wireless band channels

On the Physical layer side, the IEEE 802.15.4 defines specific RF frequencies, modulation formats, data rates and coding techniques. The 802.15.4 physical layer operates in three different unlicensed bands according to the geographical area where the system is deployed. However, spread spectrum techniques are wherever mandatory to reduce the interference level in shared unlicensed bands.

IEEE 802.15.4 specifies a total of 27 half-duplex channels across the three frequency bands and is organized as follows:

(*i*) The 868 [MHz] band: only a single channel with data rate 20 [kbps] is available; "92 [dBm] RF sensitivity required and ideal transmission range approximatively equal to 1 [km];

(*ii*) The 915 [MHz] band: ten channels with rate 40 [kbps] are available; the receiver sensitivity and the ideal transmission range are the same of the previous case;

(*iii*) The 2.4 [GHz] ISM band: sixteen channels with data rate 250 [kbps] available; minimum

(*iv*) "85 [dBm] RF sensitivity required and ideal transmission range equal to 220 [m].

9.4.2 IEEE 802.15.4 MAC Layer

IEEE 802.15.4 uses a protocol based on the CSMA/CA algorithm, which requires listening to the channel before transmitting to reduce the probability of collisions with other ongoing transmissions. IEEE 802.15.4 defines two different operational modes, namely the beacon-enabled and the non beacon-enabled, which correspond to two different channel access mechanisms. In the non beacon-enabled mode nodes use an unslotted CSMA/CA protocol to access the channel and transmit their packets. The algorithm is implemented using units of time called backoff periods. There exists a maximum number of time the node can try to access the channel. When this maximum is reached, the algorithm ends and the transmission cannot occur.

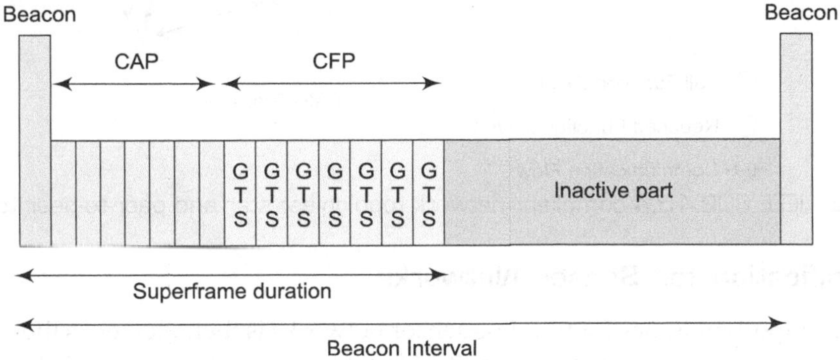

Fig. 9.4 Super frame structure

In the beacon-enabled mode, instead, the access to the channel is managed through a super frame structure as shown in Fig. 9.4, starting with a packet, and called beacon, transmitted by WPAN coordinator. The super frame may contain an inactive part, allowing nodes to go in sleeping mode, whereas the active part is divided into two parts: the Contention Access Period (CAP) and the Contention Free Period (CFP) composed of Guaranteed Time Slots (GTSs) which is optional, that can be allocated by the sink to specific nodes.

9.4.3 IEEE 802.15.4 Network Layer

To overcome the limited transmission range, multi-hop self-organizing network topologies IEEE 802.15.4 defines two types of devices: the Full Function Device (FFD) and the Reduced Function Device (RFD).

The FFD contains the complete set of MAC services and can operate as either a PAN coordinator or as a simple network device. The RFD contains a reduced set of MAC services and can operate only as a network device.

Wireless Personal Area Networks in IEEE 802.15.4 can be implemented with a star or a peer-to-peer topology, as shown in Fig. 9.5, with both full- function and reduced-function devices. Peer-to-peer topology is preferable in case a large area should be covered and latency is not a critical issue. IEEE 802.15.4 can also support other network topologies, such as cluster, mesh, and tree. Star topology is preferable in case coverage area is small and low latency is required by the application

The star topology has a single central controller, called PAN coordinator that held the communication with all other devices. Its main role is to control the network, and route the communications between two devices, if needed. The PAN coordinator can also be powered, whereas in most of the cases the other devices will be running on batteries.

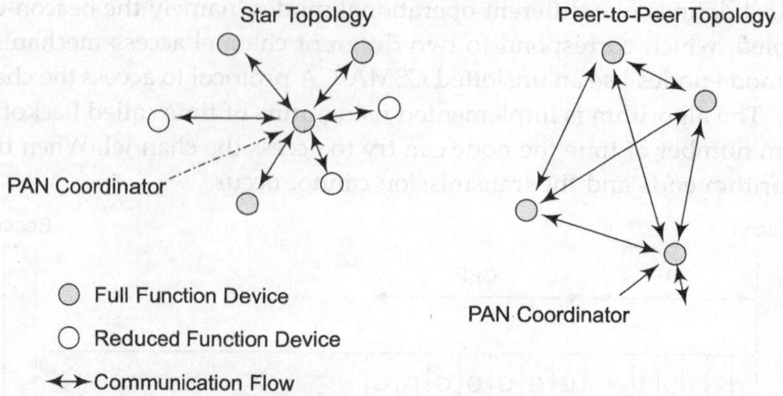

Fig. 9.5 IEEE 802.15.4-compliant network topologies: star and peer-to-peer topology

9.4.4 Specification for Sensor Networks

One of the most important issues for wireless sensor networks is their element's lifetime. In most of the cases, those networks are meant for monitoring situations, where access is hazardous or too expensive. Maintenance operations must then be avoided, and maximizing the elements lifetime goes in that way.

The other important component is the batteries that will be affected by the external conditions, but their lifetime can be extended by proper power management.

(*i*) The first possibility for saving batteries is by dividing the mote's life into active and inactive behavior. While active, the mote executes into sleep mode and it included the IEEE 802.12.4 Standard, with ûxed length.

(*ii*) Secondly possibility is to choose the appropriate sampling rate.

9.5 OVERVIEW OF ZIGBEE

ZigBee-style networks began to conceive around 1998, when many installers realized that both WiFi and Bluetooth were going to be unsuitable for many applications. The IEEE 802.15.4 standard started and was completed by May 2003. It was during that period in October 2004 the ZigBee Alliance announced members to more than 100 member companies, in 22 countries. By December 2005 membership had grown to more than 200 companies. The ZigBee Alliance announces the Standard in September 2006, known as ZigBee 2006 Specification and during the last quarter of 2007, ZigBee PRO, specification was finalized.

ZigBee defines the network and application layers above the 802.15.4. ZigBee is being promoted by the ZigBee Alliance and its main contribution is giving mesh network capabilities to 802.15.4 applications.

WLAN and Bluetooth standards have similar physical channel structure. But, there is no possibility of connection between ZigBee and WLAN (IEEE 802.11) or Bluetooth (IEEE 802.15.1) due to different specifications covered in the standards. The 802.15.4 core system consists of a radio frequency (RF) transceiver and the protocol stack, depicted in Fig. 9.6.

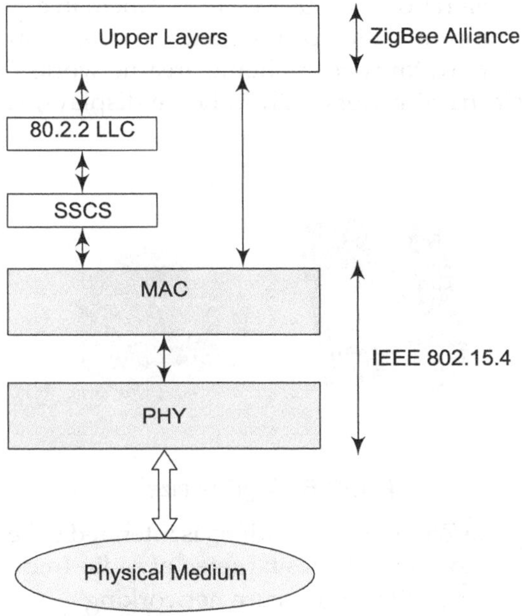

Fig. 9.6 ZigBee protocol stack

The ZigBee alliance specifies the IEEE 802.15.4 as the physical and MAC layer and is seeking to standardize higher level applications such as lighting control and HVAC monitoring. The ZigBee Alliance is an association of companies working together to enable reliable, cost-effective, low-power, wirelessly networked monitoring and control products based on an open global standard.

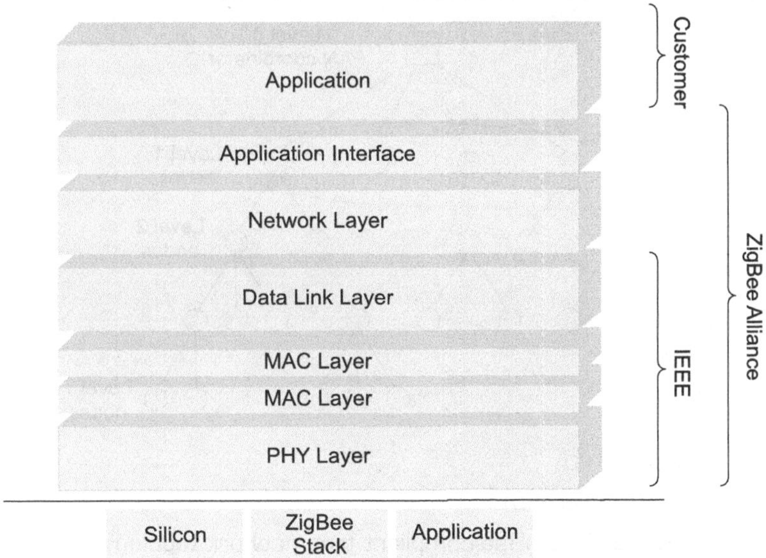

Fig. 9.7 ZigBee stack

The ZigBee network specification, to be ratified in 2004, will support both star network and hybrid star mesh networks. As can been seen in Fig. 9.7, the ZigBee alliance encompasses the IEEE802.15.4 specification and expands on the network interface. The ZigBee Alliance is a group of companies that maintain and publish the ZigBee standard.

The Fig. 9.8 shown for size reference is about 23 mm (0.9 inch) in diameter. ZigBee is a specification for a suite of high level communication protocols using small, low-power digital radio based on the IEEE 802.15.4-2003 standard for wireless home area networks (WHANs), such as wireless light switches with lamps, electrical meters with in-home-displays, consumer electronics equipment via short-range radio.

Fig. 9.8 ZigBee size

The technology defined by the ZigBee specifications is intended to be simpler and less expensive than other WPANs such as Bluetooth. ZigBee is targeted at radio-frequency (RF) applications that require a low data rate, long battery life, and secure networking.

9.6 ZIGBEE TOPOLOGY

ZigBee specifications define a beacon-enabled tree-based topology. This topology, as shown in Fig. 9.9 interpreted as a hierarchical tree where, nodes at a given level transmit data to nodes at a lower level, to reach the PAN coordinator, which is the root of the tree.

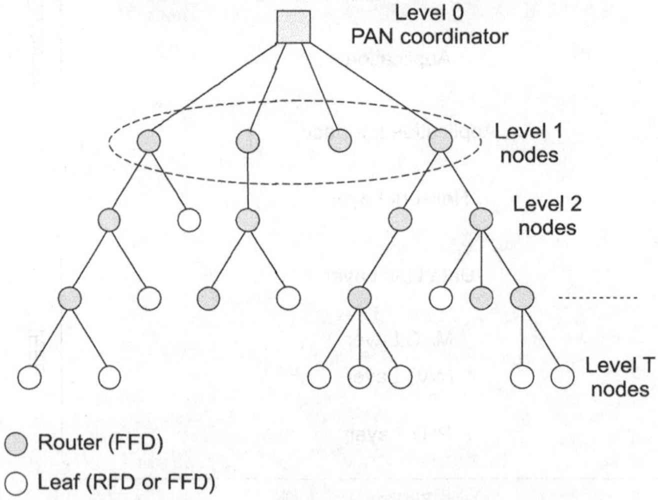

Fig. 9.9 ZigBee-compliant tree network topology

Only one device in tree assumes the role of PAN coordinator that is generally the sink of the scenario. In case of multi-sink scenario more PAN coordinators are present and a forest of disjoint trees, rooted at the PAN coordinators is established.

ZigBee defines two layers of the OSI (Open Systems Interconnection) model: the Application Layer (APL) and the Network Layer (NWL), as depicted in Fig. 9.10.

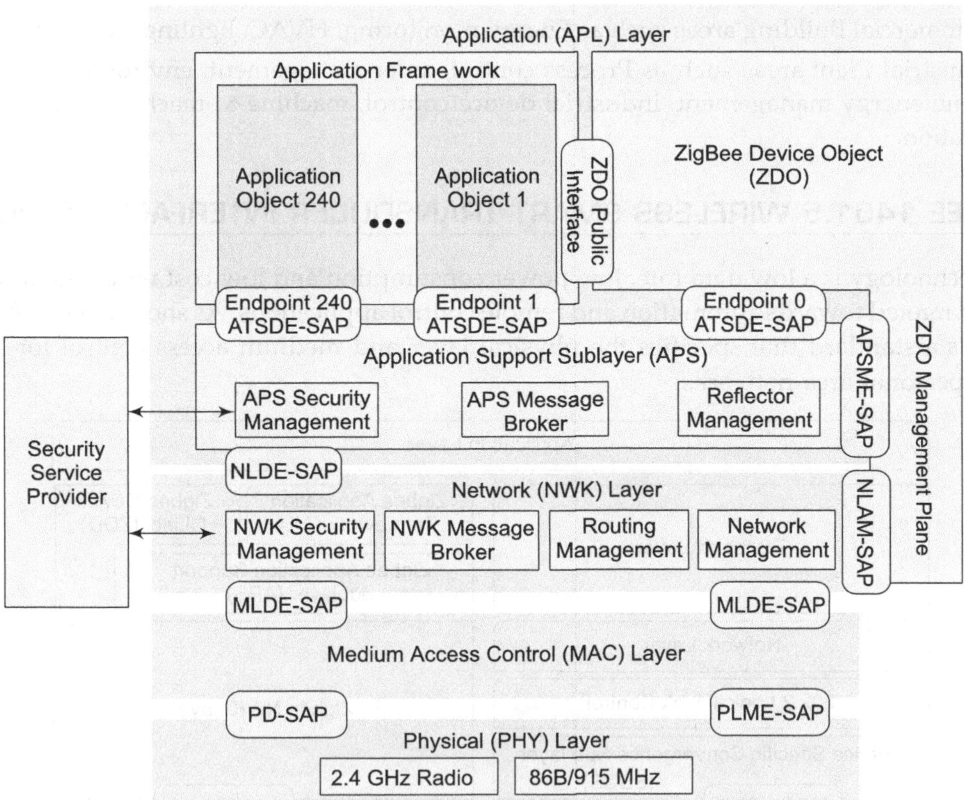

Fig. 9.10 ZigBee architecture

Each layer performs a specific set of services for the layer above. The different layers communicate through Service Access Points (SAP's). These SAPs enclose two types of entities:

(*i*) A data entity (NLDE-SAP) to provide data transmission service and

(*ii*) A management entity (NLME-SAP) providing all the management services between layers.

The ZigBee Device Object (ZDO) is also responsible for communicating information about itself and its provided services. The set of ZDOs, their configuration and functionalities form a ZigBee profile. The ZigBee profiles intent to be a uniform representation of common application scenarios. The software is designed to be easy to develop on small, inexpensive microprocessors.

9.7 APPLICATION IN ZIGBEE

Typical applications of ZigBee are as follow:

(*i*) Home Entertainment and Control areas such as Smart lighting, advanced temperature control, safety and security, movies and music.

(*ii*) Home Awareness areas such as Water sensors, power sensors, energy monitoring, smoke and fire detectors, smart appliances and access sensors.

(*iii*) Mobile Services areas such as *m*-payment, m-monitoring and control, *m*-security and access control, *m*-healthcare and tele-assistance.

(*iv*) Commercial Building areas such as Energy monitoring, HVAC, lighting, access control.

(*v*) Industrial Plant areas such as Process control, asset management, environmental management, energy management, industrial device control, machine-to-machine (M2M) communication.

9.8 IEEE 1451.5 WIRELESS SMART TRANSDUCER INTERFACE STANDARDS

ZigBee technology is a low data rate, low power consumption and low cost wire-less networking protocol targeted towards automation and remote control applications. As shown in Fig. 9.11, IEEE 802.15.4 is a standard that specifies the physical layer and medium access control for low-rate wireless personal area networks.

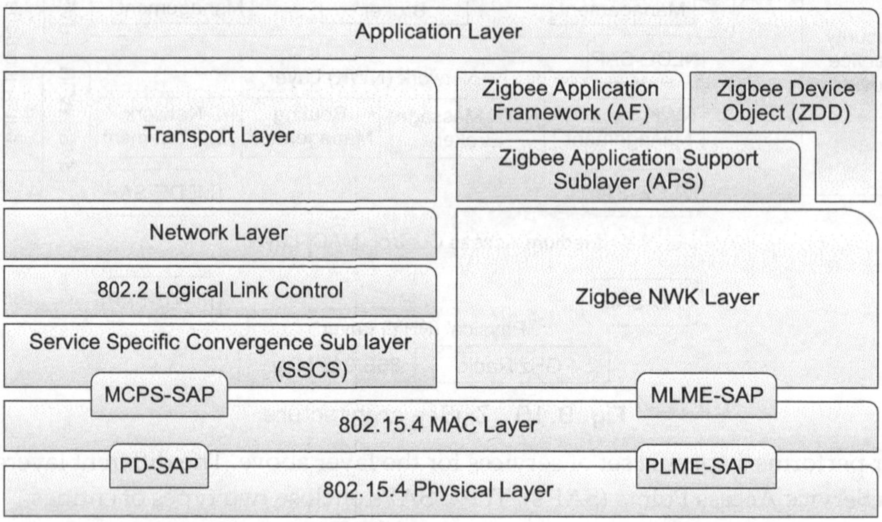

Fig. 9.11 Comparison of the IEEE 802.15.4 and ZigBee standards OSI layers

IEEE 802.15.4 standard intends to offer the fundamental lower network layers and focuses on low-cost, low-speed ubiquitous communication between devices.

IEEE 802.15.4 focuses on the lower two layers of the OSI protocol stack, whereas ZigBee Alliance aims to provide the upper layers for interoperable data net working, security services and an extremely wide range of wireless home and building control solutions. The goal is to provide as well interoperability, compliance testing, and marketing of the standard and advanced engineering solutions for the evolution of the standard.

The development of wireless sensor network technology and standards is not being driven solely by the communications interface. Manufacturers, implementers, as well as users of sensors have expressed a desire for wireless connectivity. They were finding it difficult to produce transducers compliant with the increasingly large number of network communication protocols. The solution was the development of a smart transducer interface, which is a single communication protocol usable by all sensors; this need has been recognized by the IEEE working bodies, and has led to the formation of the IEEE 1451.5 Wireless Sensor Working Group which became IEEE 1451. Fig. 9.12 represents IEEE 1451.5 smart transducer standards.

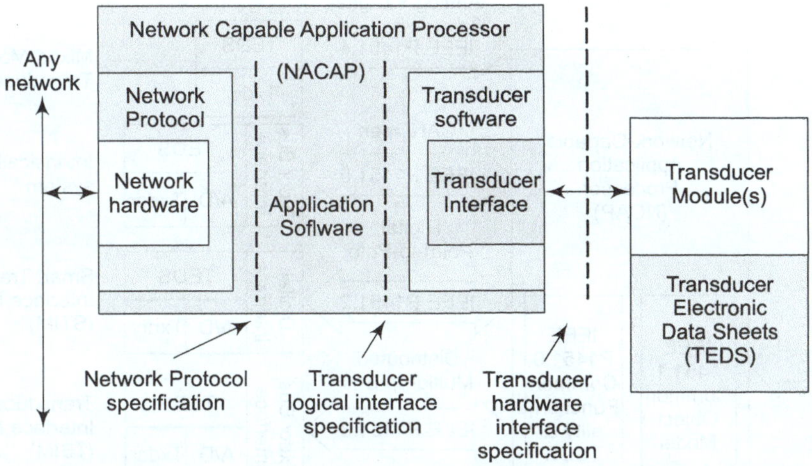

Fig. 9.12 IEEE 1451.5 smart transducer standards

IEEE 1451 is a set of Smart transducer interface standards developed by the IEEE Instrumentation and Measurement Society's Sensor Technology Technical Committee that describe a set of open, common, network-independent communication interfaces for connecting transducers (sensors or actuators) to microprocessors, instrumentation systems, and control/field networks. One of the key elements of these standards is the definition of TEDS (Transducer Electronic Data Sheet) for each transducer.

The TEDS, is a memory device that stores transducer identification, calibration, correction data, and manufacturer-related information. The objective of the IEEE 1451 family of standards is to allow the access of transducer data through a common set of interfaces whether the transducers are connected to systems or networks via a wired or wireless means.

The 1451 family of standards is:

(*i*) 1451.0-2007 IEEE Standard is for a Smart Transducer Interface for Sensors and Actuators – Common Functions, Communication Protocols, and Transducer Electronic Data Sheet (TEDS) Formats

(*ii*) 1451.1-1999 IEEE Standard is for a Smart Transducer Interface for Sensors and Actuators – Network Capable Application Processor Information Model

(*iii*) 1451.2-1997 IEEE Standard is for a Smart Transducer Interface for Sensors and Actuators – Transducer to Microprocessor Communication Protocols & TEDS Formats

(*iv*) 1451.3-2003 IEEE Standard is for a Smart Transducer Interface for Sensors and Actuators – Digital Communication & TEDS Formats for Distributed Multidrop Systems

(*v*) 1451.4-2004 IEEE Standard is for a Smart Transducer Interface for Sensors and Actuators – Mixed-Mode Communication Protocols & TEDS Formats

(*vi*) 1451.5-2007 IEEE Standard is for a Smart Transducer Interface for Sensors and Actuators – Wireless Communication Protocols & Transducer Electronic Data Sheet (TEDS) Formats

(*vii*) 1451.7-2010 IEEE Standard is for a Smart Transducer Interface for Sensors and Actuators – Transducers to Radio Frequency Identification (RFID) Systems Communication Protocols and Transducer Electronic Data Sheet Formats

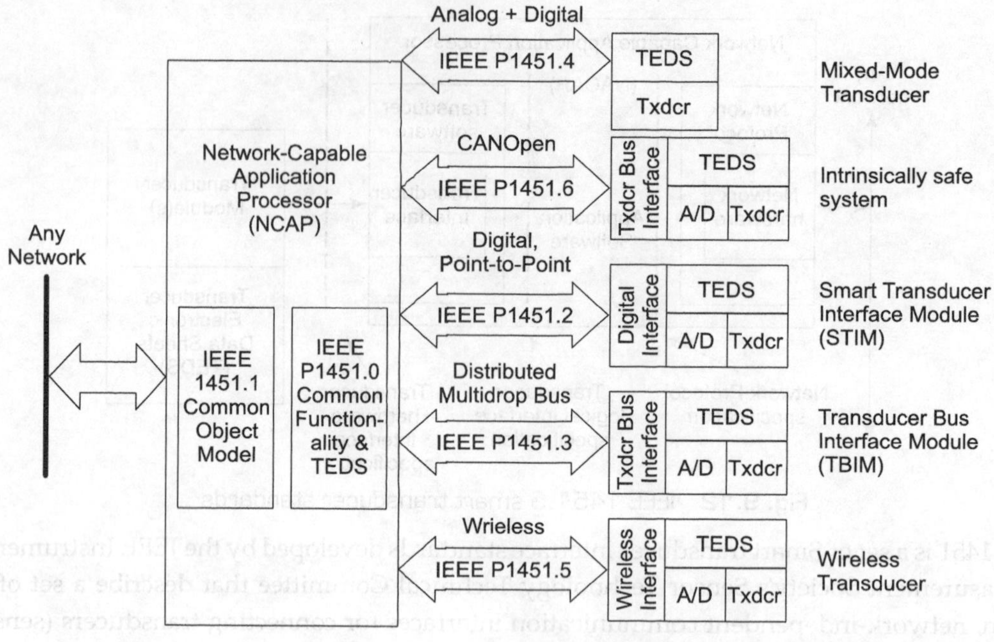

Fig. 9.13 IEEE 1451 family of smart transducer interface standards

The shown Fig. 9.13 represents IEEE 1451 Family of Smart Transducer Interface Standards The proposed IEEE 1451.5 Wireless Sensor Standard the consensus to adopt the popular wireless communication protocols such as

(*i*) Bluetooth/802.15.1

(*ii*) ZigBee/802.15.4

(*iii*) Wi-Fi/802.11 as the P1451.5 physical layers.

SUMMARY

In this chapter discussed a work on the IEEE 802.15.4/ZigBee protocol stack. The topic focus on various IEEE standards and also mainly on IEEE 802.15.4 and ZigBee protocols as a baseline to easier, faster and widespread development, deployment and adoption.

QUESTIONS

1. What are the different wireless standards?
2. Draw the IEEE 802 Main standards.
3. Explain the IEEE 802.15.4 architecture.
4. Broadly explain the IEEE 802.15.4 layers.
5. Explain the overview of ZigBee.
6. Illustrates the ZigBee protocol stack.
7. Draw and explain the ZigBee architecture.
8. Discuss the IEEE 1451.5 smart transducer standards.

BIBLIOGRAPHY

- G. J. Pottie and W. J. Kaiser. "Wireless integrated network sensors". Communications of the ACM, 43(5), May 2000.
- Koubaa and M. Alves, "OPNET simulator for IEEE 802.15.4 protocol, release 1.0, http://www.open-zb.net/download," 2006.
- K. Romer and M. Friedemann, "The Design Space of Wireless Sensor Networks," IEEE Wireless Communications, vol. 11, pp. 54-61, 2004.
- G. Montenegro, N. Kushalnagar, J. Hui and D. Culler. Transmission of IPv6 Packets over IEEE 802.14.4 Networks Internet-Draft draft-ietf-6lowpan-format-13 (work in progress), April 2007.
- Archana Bharathidasan, Vijay Anand Sai Ponduru, Sensor Networks: An Overview. Technical report, University of California: IEEE Potentials, 2003.
- IEEE, IEEE Standard 802.15.4, pdf, 2003
- ZigBee Alliance, http://www.caba.org/standard/zigbee.html
- IEEE 802.15.4a Standard: Wireless MAC and PHY Specifications for Low-Rate Wireless Personal Area Networks (LR-WPANs): Amendment to add alternate PHY (Draft). IEEE: Piscataway, NJ, USA, 2006.

10
Test-bed in WSN

10.1 INTRODUCTION

Recent advancements in wireless communications and micro sensing embedded technologies are enabling the real deployment of Wireless Sensor Networks (WSN). It is a highly distributed network of small, lightweight wireless nodes. Recently, WSNs have been used in many promising applications including habitat monitoring, military target tracking, natural disaster relief, and health monitoring.

A WSN is a special kind of an ad-hoc network composed by hundreds, even thousands, of wireless sensor nodes. Each wireless sensor node is equipped with a microprocessor for data processing, radio chip for wireless communication and sensor board for sensing some physical phenomena. The advent of Wireless Sensor Network technologies is paving the way new ubiquitous computing applications, some of them with critical requirements. However, WSNs are resource constrained with a limited energy lifetime, slow computation, small memory, and limited communication capabilities. WSN consists of a base station (or "gateway") that can communicate with a number of sensors via a radio network as shown in Fig. 10.1.

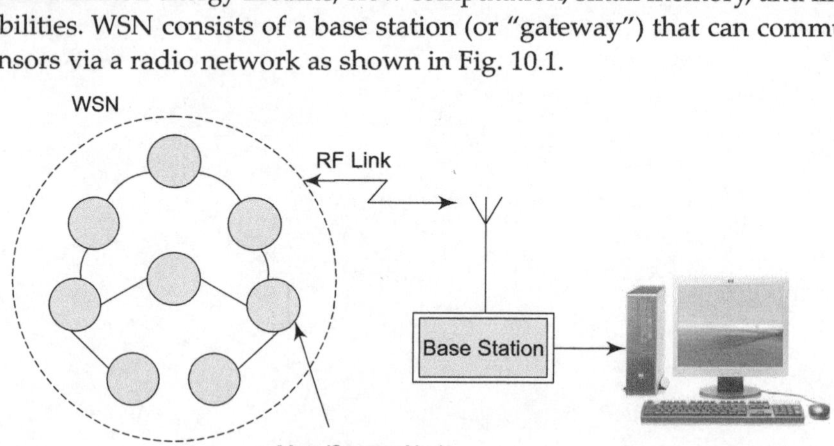

Fig. 10.1 Basic WSN architecture

Data is collected at the wireless sensor nodes or motes (comprising sensors, data acquisition circuit, and processor and RF transceiver), compressed, and transmitted to the gateway directly or, if required, uses other wireless sensor nodes to forward data to the gateway. The transmitted data is then presented to the system by the gateway connection.

In this chapter, discuss the following various types of experimental study on test-bed of a sensor devices to simulate and to implement the solutions for implementation and analysis.

10.2 A TEST-BED APPLICATION OF ANALOG AND DIGITAL SENSORS

The most widely accepted hardware platform is the UC Berkeley mote platform in the WSN. The mote platform uses an 8-bit microprocessor, such as ATMEGA 128, ATMEL 8535, or Motorola HCS08, with a 4MHz CPU. Typically, the motes have 4kB RAM, which is used for holding run-time state of the program with an additional flash memory storage space up to 512 kB.

10.2.1 Experimental Lab Connection

Figure 10.2 shows the test-bed of analog and digital sensor schematic diagram of hardware connections for the wireless senor network.

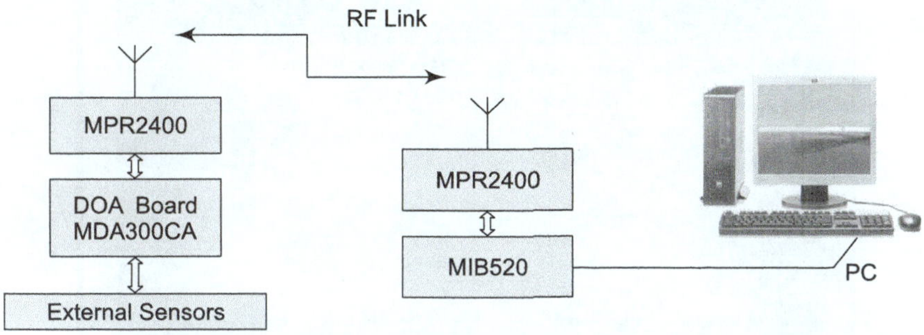

Fig. 10.2 Schematic diagram of hardware connections

The actual experimental setup for acquiring data from externally connected potentiometer and door and window digital sensors is denoted in the Fig. 10.3.

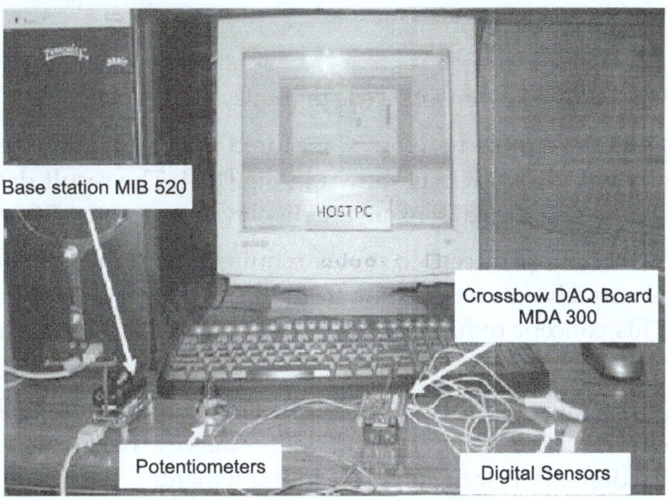

Fig. 10.3 Experimental setup

The method of this experiment is based on the following steps:

1. Plug the Mote (MPR2400) into the programming board MIB520,
2. Select Tools > shell and type make micaz reinstall, <node Id> mib520,com5. The node ID is selected between 1 and 65534 for the mote that function as the data acquisition board and node ID 0 for the mote that functions as the base station.

The steps required to compile and install the XMeshBase application onto the base station node are:

1. Select the XMeshBase.nc file in Programmer's Notepad,
2. Select Tools > shell and type make micaz install,0 mib520,com5.

After the completion of programming MDA 300 DAQ mote, the base station mote is plugged into the programming board and the data acquisition node mote is plugged into the MDA 300CA DAQ board as shown in Fig. 10.2.

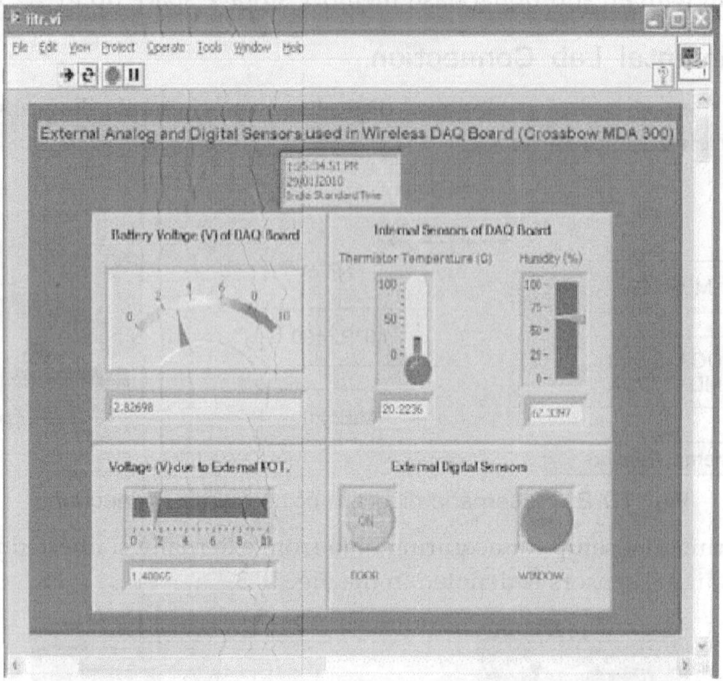

Fig. 10.4 Experimental results displayed on labVIEW GUI

The information collected from the network was stored in a local database and made available for displaying through a LabVIEW GUI as shown in Fig. 10.4. The result data acquired from the externally connected analog and digital sensors are displayed in on host PC by running LabVIEW.

The integration of custom transducers to motes required the implementation of signal conditioning algorithms to accurately extract the data from the devices and apply it to the data acquisition board channels. This concept reduces the barrier of connecting different analog and digital sensors in wireless sensor networks.

The lab setup gives the idea about the possibilities of adapting external analog and digital sensors for use with wireless sensor networks and determines the hardware and software requirements for making such adaptations more efficient to implement.

10.3 A TEST-BED APPLICATION OF WORMHOLE ATTACK IN WSN

WSNs are vulnerable to several attacks. One major attack that is known as a wormhole attack where an attacker records a packet at one location in the network, tunnels the packet to another location, and then replays it at another part of the network. The wormhole attack places the attacker in a very powerful position, allowing him to gain unauthorized access, disrupt routing, or perform a Denial-of-Service (DoS) attack. The current solutions for wormhole attacks are evaluated by running the proposed techniques in different simulations. There is no real deployment for any of these solutions. Some of the methods are crisp implementation of the wormhole attack in WSNs; BANAID consists of Mica2, Stargate sensor devices.

10.3.1 Experimental Lab Setup

The implementation of BANAID evaluating different solutions for the wormhole attack in the wireless sensor networks applications.

The worm attack solutions, using Packet Leashes is shown below. BANAID consists of several Mica2 motes. The MICA2 motes as shown in Fig. 10.5 come in three types according to the RF frequency bands: MPR400 (915 MHz), MPR410 (433 MHz), and MPR420 (315 MHz).

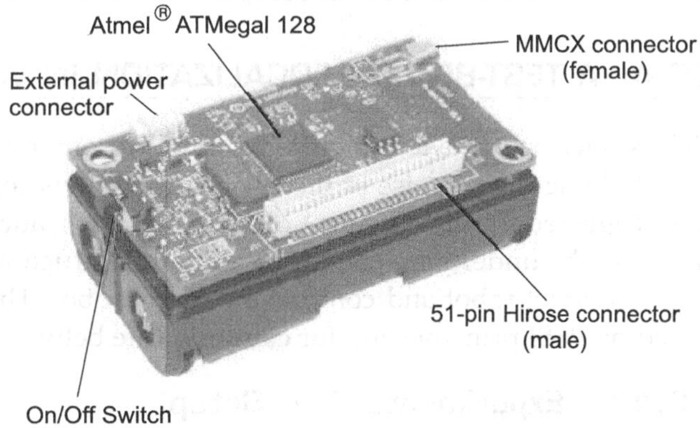

Fig. 10.5 MICA2 mote without an antenna

The motes use the Chipcon CC1000, FSK modulated radio. All types utilize a powerful Atmega128L microcontroller and a frequency tunable radio with extended range. The MPR4x0 and MPR5x0 radios are compatible and can communicate with each other. Figurel shows a sample of MICA2 mote. The current version of Mica2 uses a 16 bit, 8MHz Texas Instruments, 1024 KB external flash, and is powered by two AA batteries.

The MICA2 mote can be reprogrammed using an external board called MIB510 Serial Interface Board as shown in Fig. 10.6.

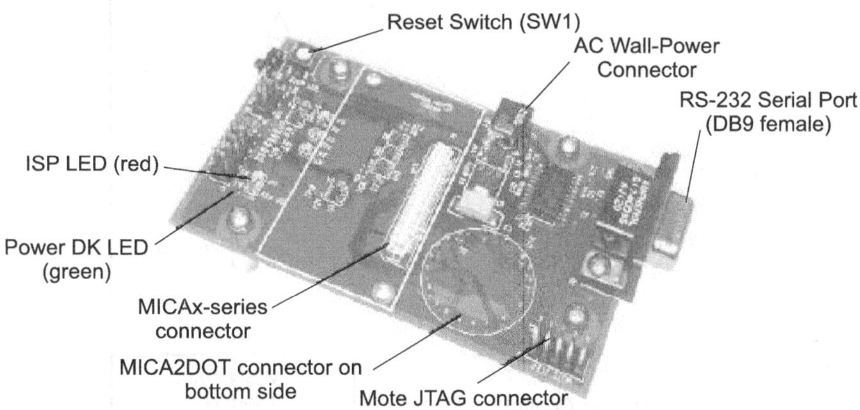

Fig. 10.6 MIB510 serial interface board

The board is a multi-purpose interface board used with MICA, MICA2, MicaZ, and MICA2DOT motes family. It provides an interface for a RS-232 mote serial port and reprogramming port. The MIB510 serial interface board is used to program the MICA2 mote. This board has the PC connection capability using the RS-232 serial port. Programming the motes requires a special operating system called TinyOS1, which should be installed in the PC.

MICA2 motes use a special operating system called TinyOS, which is used for wireless sensor nodes. This operating system is an open-source event-driven operating system designed for wireless embedded sensor networks. It features a component based architecture which enables rapid innovation and implementation while minimizing code size as required by severe memory constraints inherent in sensor networks.

Tiny OS's component library includes network protocols, distributed services, sensor drivers, and data acquisition tools - all of which can be used as-is or can be further refined for a customized application. Tiny OS have been implemented in a C-language called nesC 2. Programs written in nesC language are built out of components, which are wired to form entire programs.

10.4 A TEST-BED ON LOCALIZATION IN WSN

WSNs offer new ways to monitor/control our environment in a continuous and almost invisible way, holding the promise of many new ubiquitous computing applications. The theoretical findings, triggered the development of a test-bed application, since it imposes stringent timing requirements to the underlying communication infrastructure. The control station detects, localizes and tracks a target robot and controls the target robot. The WSN motes are used for the localization based on RSSI principle and for communicate between the different entities involved.

10.4.1 Experimental Lab Setup

The overall objective of the application is to detect, localize and rescue a target entity, within a certain region covered by a WSN deployment. Mobile robots are currently being used to act as target and rescuer entities. The target robot is supposed to be in distress or to be an intruder. A control station instruct the rescuer and navigate towards the target robot, until it gets close enough to it as shown in Fig. 10.7.

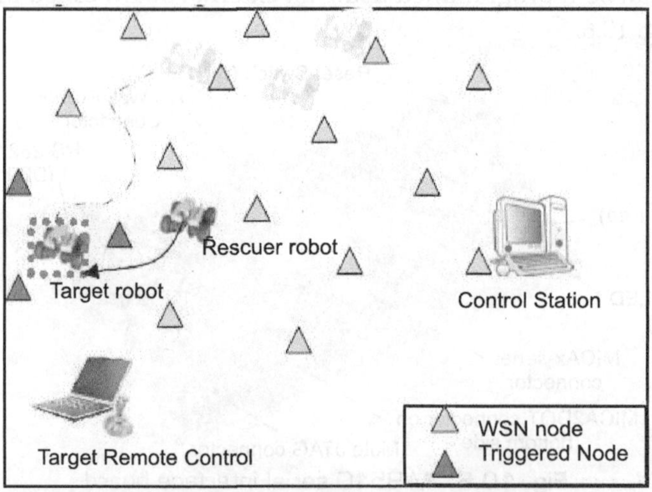

Fig. 10.7 Snapshot of the application

The target robot movement is remotely controlled by an operator, using a joystick. A WSN node mounted on top sends periodic messages to signal its presence, which are then relayed by the WSN to the Control Station. The Control Station implements a 3D virtual display of rescuer and target robots status. The localization of target is based on messages sent from the rescuer robot displaying status such as: position, heading, mission status, waypoints. The Control Station then computes the target robot location and informs the rescuer robot that will immediately initiate its mission by moving towards the last known position of the target robot in an autonomous fashion. This process is repeated until the rescuer robot is close enough to the target robot.

Fig. 10.8 presents the current test-bed deployment, showing a rescue mission in progress. The WSN nodes are MicaZ motes, featuring an 8-bit microcontroller with 128KB of in-system programmable memory. In order to interface the rescuer robot and the WSN, a MIB510 interface board.

Fig. 10.8 Picture of the current indoor deployment

The MicaZ nodes run TinyOS, an open-source event-driven operating system. The mobile robotic platform used in the application is the WifiBot. The system architecture is build around a double Ethernet-I²C bus and a CPU-4G Access Cube features an IEEE 802.11 access point. A hardware platform dedicated to Wireless LAN Mesh Routing acting as an interface between the two.

The study focus on the theoretical and practical aspects related to the implementation of a localization mechanism based on radio signal strength measurements. This practical experience on the technology use test-bed is of paramount importance for future developments.

Future work test-bed application will show further scope for researcher and the extension of the WSN deployment, enlarging the rescuer and target robot teams and the integration of IEEE802.15.4/ZigBee protocol stack in the MicaZ motes for supporting multi-hop real-time communications in the WSN.

10.5 ENERGY OPTIMIZATION APPROACH ON MICAZ MOTE

The implementation of test-bed hardware using the low power MicaZ mote represents the latest generation of Berkeley motes, and is commercialized by Crossbow. It is built around an ATMega128L microprocessor, and features a CC2420 802.15.4-compliant ZigBee-ready radio. IEEE 802.15.4 is a standard for low-rate, wireless personal area networks which provides specification for the physical and the MAC layer. At the physical layer, it defines a low-power spread spectrum radio operating at 2.4GHz with a bit rate of 250kb per second. ZigBee is a collection of high level communication protocols built on top of the IEEE 802.15.4 MAC layer.

Fig. 10.9 MicaZ mote test-platform

The scheme is implemented as shown in Fig. 10.9 with the 8 MicaZ motes, operation system TinyOS is placed in a lab environment in an arrangement shown in Fig.6. Mote 1 and mote 2 are source nodes which generates a data packet in every 5 seconds. Mote 0 is the sink mode set an MIB600 gateway to receive data from source nodes. Fresh batteries are used. To evaluate the runtime performance of EOA, EOA compare with against MicaZ mote's default MAC protocol which is similar to B-MAC, but has the Low Power Listening functionality. Besides, this MAC protocol associates with a routing protocol named as DSDV which has been implemented in TinyOS. Entire experiment conducted runs for a 72-hour to demonstrate the advantage of EOA.

SUMMARY

This chapter outlines focus on the theoretical and practical aspects related to the implementation of a localization mechanism based on radio signal strength measurements. This practical experience on the technology under use in the test-bed is of paramount importance for future developments. Nevertheless, some improvements regarding the application design were identified, like the computation of the target robot localization in the WSN. Future there is a discussion on test-bed application that address issues of the WSN deployment and the integration of IEEE 802.15.4/ZigBee protocol stack in the MicaZ motes, for supporting multi-hop real-time communications in the WSN.

QUESTIONS

1. Why there is the requirement of test-bed in WSN?
2. Explain in detail the test-bed of analog and digital sensors.
3. Discuss the test-bed application of wormhole attack in WSN.
4. Brief out the test-bed on localization in WSN.

BIBLIOGRAPHY

- R. Severino, M. Alves, "On a Test-bed Application for the ART-WiSe Framework", IPP-HURRAY Technical Report TR-061103, November 2006.

- "Lab VIEW Basics I Course Manual" National Instrument, Document Number 320628G-01, September 2000.
- M. Alves, A. Koubaa, A. Cunha, R. Severino, E. Lomba, "On the Development of a Test-Bed Application for the ART-WiSe Architecture", WiP Session of the 18th Euromicro Conference on Real-Time Systems (ECRTS'06), Germany, July, 2006.
- A Wireless Sensor Networks Test-bed for the Wormhole Attack Hani Alzaid and Suhail Abanmi, International Journal of Digital Content Technology and its Applications Volume 3, Number 3, September 2009.
- Crossbow Technology, Inc. Stargate Developer Manual. San Jose, CA, USA, May 30-31, 2005. This Stargate Developer Manual is retrieved 10th of May 2009.
- Ian F. Akyildiz, Weilian Su, Yogesh Sankarasubramaniam, and Erdal Cayirci. Wireless sensor networks: a survey. Computer Networks, 38(4):393-422, 2002.
- A Test-Bed Application of Analog and Digital Sensor for WSN, Mrutyunjay Rout, H.K Verma, Indian Institute of Technology Roorkee (IITR).
- "MPR-MIB User's Manual." Crossbow Technology, Document Number 7430- 0021-08, Rev. A, June 2007.

11 Security in WSN

11.1 INTRODUCTION

Wireless Sensor Network (WSN) is an emerging technology that shows great promise for various futuristic applications both for mass public and military. The sensing technology combine with processing power and wireless communication that exploited in abundance in future. The enclosure of wireless communication technology also incurs various types of security threats. Due to inherent limitations in wireless sensor networks, security is a crucial issue. Lot of research work is carrying on the security related issues and challenges in wireless sensor networks. The research in WSN security is progressing at tremendous pace, and no comprehensive document lists the security issues and the threat models which pose unique threats to the wireless sensor networks.

Wireless Sensor Network (WSN) is a hundreds or thousands of low-power, low-cost nodes and small size sensor nodes. These sensor nodes can sense, process and communicate information among them. It is a heterogeneous system combining tiny sensors and actuators with general-purpose computing elements. There are several applications that take place in the WSN where several processing data plays the vital information exchange. These network processing to reduce large streams of raw data into useful aggregated information. So, the protection of this sensor data is critical.

Sensor networks pose unique challenges where, traditional security techniques used in traditional networks cannot be applied directly for wireless sensor network.

- First, to make sensor networks economically viable, sensor devices of limited energy -consumption, computation, and communication capabilities.
- Second, unlike traditional networks, sensor nodes deployment can be easily accessible.
- And third, sensor networks work together closely with their physical environments and with people, posing new security problems.

Wireless stations, or nodes, communicate over a wireless medium Networks operating under infrastructure mode e.g., 802.11, 802.16, Cellular networks. Networks operating architecture security threats that are imminent due to the open nature of communication its two main issues:

1. Authentication and privacy and
2. Other serious issues: denial-of-service.

For some sensor network applications security is crucial factor as they are deployed in hostile environments with active intelligent opposition. One obvious important security reliant applications are:

- Disasters: In many disaster scenarios, especially those induced by terrorist activities, it may be necessary to protect the location of casualties from unauthorized disclosure
- Public Safety: In applications where chemical, biological or other environmental threats are monitored, it is vital that the availability of the network is never threatened. Attacks causing false alarms and it may lead to panic responses or even worse total disregard for the signals.
- Home Healthcare: In such applications, privacy protection is essential. Only authorized users should be able to query and monitor the network.

11.2 WSN SECURITY STUDY

Wireless sensor devices that are employed for security applications have several functionalities. The first one is the distributed detection of the presence of a target, and the estimation of parameters of interest. The target detection, estimation and tracking efforts may or may not be collaborative.

The second task involves wireless networking to organize and carry information. As shown in Fig. 11.1 such threat can be interacted at the node end. Here, the user communicates to the sink node in a wireless ad-hoc network which is a treat the network from intruders. The attacker can break the security of gateway to access the further route of nodes information.

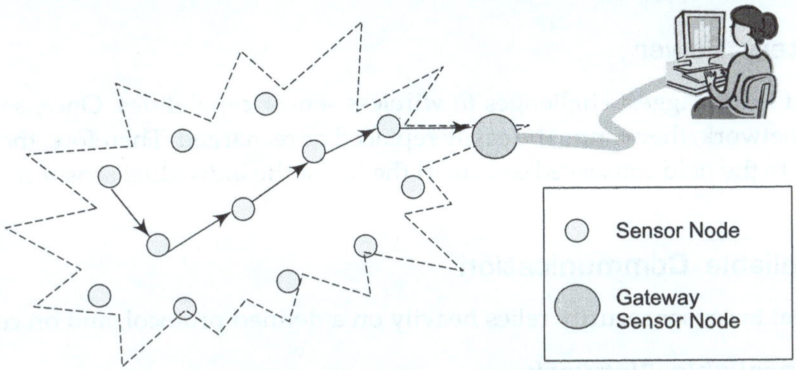

Fig. 11.1 Nodes spreading and access for security threats

The topology of wireless networks is the basis of data transmitting should have node-failure tolerance ability when disposed in hostile region. When the sensors are unattended and it will be failure by fault, intrusion, and energy exhausting. The sensor nodes are limited by the bandwidth, memory, and computational capability which will make the traditional asymmetrical cryptographic methods unavailable to protect the intrusion. Further, the sensors that are unattended are doomed to failure, because the battery will be exhausted.

The security is important factor in WSNs because of cost effective and practical way to go deploying sensor networks. The threats for military application pose different challenges as compared to traditional networks. Hence, different mechanisms must be brought about with enormous research potential.

The existing security mechanisms are inadequate for WSN, and researchers are finding opportunity to properly address sensor network security. Some of the Background-Problem Statement is: Network Assumptions, Insecure links - non denial of eavesdropping, Replay and injection of packets, Non tamper resistant, Trust Requirements, Base station and aggregation point trustworthiness.

Some other areas are outsider vs. insider attacks, Mote-class vs. laptop-class attackers, security Goals, Integrity, authentication, freshness and confidentiality.

11.3 IMPEDIMENT OF WIRELESS SENSOR NETWORK SECURITY

A wireless sensor network is a special network which has many constraints compared to a traditional computer network. Due to these constraints it is difficult to directly employ the existing security approaches to the area of wireless sensor networks. Therefore, to develop useful security mechanisms is a challenging task. Some of the obstacles that hinder the security of WSN are mentioned below.

11.3.1 Very Limited Resources

All security approaches require a certain amount of resources are very limited in a tiny wireless sensor for the implementation, including data memory, code space, and energy to power the sensor. For example, one common sensor type (TelosB) has an 16-bit, 8 MHz RISC CPU with only 10K RAM, 48K program memory, and 1024K flash storage.

The total code space of TinyOS, the defacto standard operating system for wireless sensors, is approximately 4K, and the core scheduler occupies only 178 bytes. Therefore, the code size for the all security related code must also be small.

11.3.2 Battery Power

Power constraint is the biggest challenges to wireless sensor capabilities. Once, sensor nodes are deployed in the network, they cannot be easily replaced or recharged. Therefore, the battery charge taken with them to the field conserved to extend the life of the individual sensor node in the entire sensor network.

11.3.3 Unreliable Communication

Some of the threat to sensor security relies heavily on a defined protocol, and on communication.

11.3.3.1 Unreliable Network

The packet based routing of the sensor network is connectionless and thus inherently unreliable. Packets may get damaged due to communication channel error rate that results due to missing packets. More importantly, if the protocol lacks the appropriate error it is possible to lose critical security packets.

11.3.3.2 Conflicts

The channel reliability of the communication depends due to the broadcast nature of the wireless sensor network. If packets are crowded in high density traffic its transfer will fail and lead to major conflict problem.

11.3.3.3 Latency

The latency of the network can cause due to multi-hop routing, network congestion, and node processing thus making it difficult to achieve synchronization among sensor nodes. The synchronization issues can be critical to sensor security reports.

11.3.4 Unattended Operation

The security threat is very much prone when the sensor nodes are left unattended for long periods of time depending on the function of the sensor network. There are three main reasons to unattended sensor nodes:

11.3.4.1 Exposure to Physical Attacks

When the sensor is deployed in an environment open to adversaries, bad weather, and so on it may likely that a sensor suffers a physical attack to withstand.

11.3.4.2 Managed Remotely

Remote management of a sensor network makes it virtually impossible to detect physical tampering and maintenance issues on battery replacement. In such a case, the node may have to keep physical once deployed.

11.3.4.3 No Central Management Point

A sensor network should be a distributed network without a central management point. This will increase the vitality of the sensor network.

11.4 THREATS IN LAYERS OF WSN

Many sensor network deployments are security sensitive and attack from real-world in the form of hardware failures, bugs, resource exhaustion, malicious attacks and environmental conditions that can diminish or even eliminate a networks capacity to perform as expected. The layered network architecture as shown in Fig. 11.2 of sensor networks that specify DoS vulnerabilities to the first four layers is stated below such conditions are defined as Denial of Service (DoS) attacks. Following shows the attach relations on each layers.

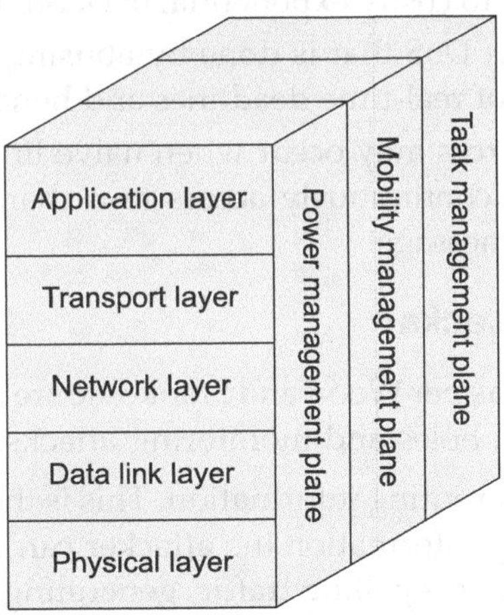

Fig. 11.2 Sensor node protocol stack

11.4.1 Physical Layer Attacks

Jamming and tampering are the most common attacks to the physical layer of a WSN.

(*a*) Jamming means it interferes with the radio frequencies to put a considerable amount of the nodes out of order. It blocks the entire network that constitutes complete DoS. In such jamming attacks jammed nodes attempt to inform neighboring nodes about the base station as shown in Fig. 11.3. Neighboring nodes can also assume a jamming attack if they observe change in the neighboring background noise.

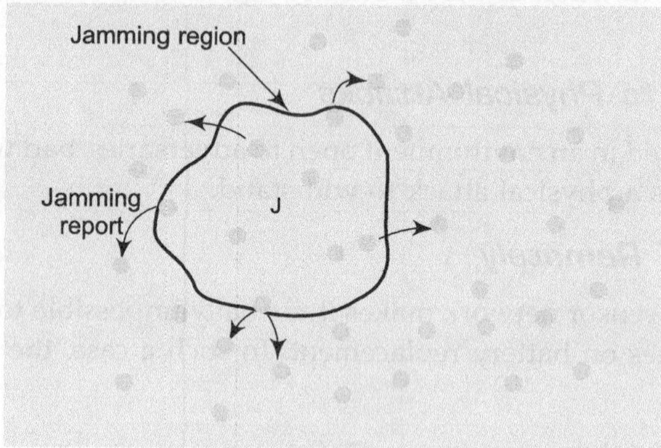

Fig. 11.3 Structure of jamming attacks

(*b*) A tampering attacker may damage a sensor, if they are part of a network to replace the entire node or part of its hardware or even electronically interrogate the nodes to gain access to sensitive information, such as shared cryptographic keys and to access higher communication layers.

11.4.2 Data Link Layer Attacks

Collisions, unfairness or exhaustion attacks can be launched against the data link layer frames of a sensor network.

(*a*) Collisions are a type of link layer jamming. If an attacker can corrupt an octet of transmission such that a checksum mismatch occurs, then the entire packet can be disrupted. Corrupted ACK messages usually lead to costly exponential back off in some MAC protocols.

(*b*) Unfairness is a weaker form DoS that is done by abusing MAC priority schemes. Such an attack usually leads to loss of real-time deadlines and hence degradation of service.

(*c*) Exhaustion of battery resources may occur when naive link layer implementations attempt repeated retransmission or continuously access to a channel, forcing its neighbor's to respond with a clear to send message.

11.4.3 Network Layer Attacks

The network layer routing attacks as per Wood and Stankovic are neglect, greed, homing, misdirection, authorization, probing, black holes and monitoring attacks.

(*a*) Spoofed, altered or replayed routing information: This is the most direct attack. By spoofing, altering or replaying routing information the attacker can complicate the network by creating routing loops, attracting or repelling traffic, generating false error messages, shortening or extending source routes or partitioning the network.

(*b*) Selective Forwarding: In such an attack the adversary includes data flow path of interest. They forward certain packets and drop portion causing a sort of black hole. Such attacks are much harder to detect than black hole attacks.

(*c*) Sinkhole Attacks: The goal of a sinkhole attack is to lure traffic to a malicious part of the network. Such attacks are usually the launching block for other attacks such as selective forwarding. Sinkholes work by making a compromised node attractive to its neighbor's by advertising high quality routes *i.e.*, low latency routes. Fooled neighbors will then forward all their data destined to the base station to the lying node.

(*d*) The Sybil Attack: The Sybil attack targets fault tolerant schemes such as distributed storage, disparity, and multipath routing and topology maintenance. This is done by having a malicious node present multiple identities to the network. This attack is especially confusing to geographic routing protocols as the adversary appears to be in multiple locations at once.

(*e*) Wormholes: In these attacks the adversary tunnels messages received in one part of the network over a low latency link, to another part of the network where the messages are then replayed. Wormholes often convince distant nodes that they are neighbor are leading to quick exhaustion of their energy resources. The simplest occurrence of this attack is to have a malicious node forwarding data between two legitimate nodes.

(*f*) Hello flood attacks: This attack similar to type of broadcast wormhole. In many routing protocols, nodes broadcast hello messages to announce their presence to their neighbor's as shown in Fig. 11.4. An attacker with a high powered antenna can convince every node in the network that it is their neighbor.

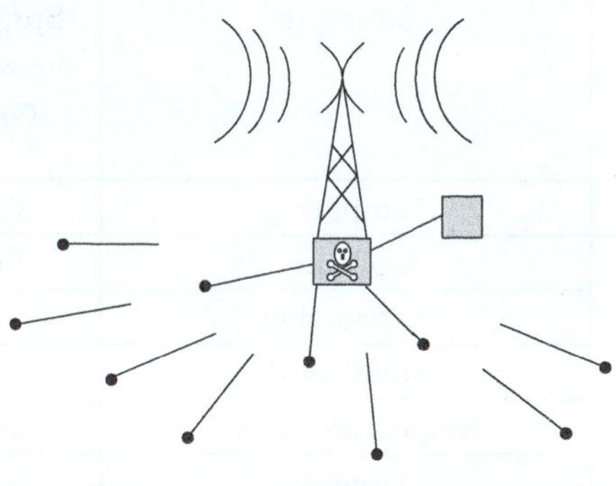

Fig. 11.4 Hello flood attack

A node receiving such a message can assume that the node that sent the message is within its range. Nodes at a large distance from the attacker will be sending their void messages leaving the network in a state of confusion.

(*g*) Acknowledgement Spoofing: This is to provide fake information will lead to action is susceptible to acknowledgments spoofing. Here the attacker spoofs acknowledgement convincing the sender that a weak link may be strong or a dead node is alive. This results in packets being lost when travelling along such links.

Due to broadcast nature of communication there are chances of packet dropping. It send lot of RREQ messages but never use the routes and behave as broadcast nature of communication as shown in Fig. 11.5.

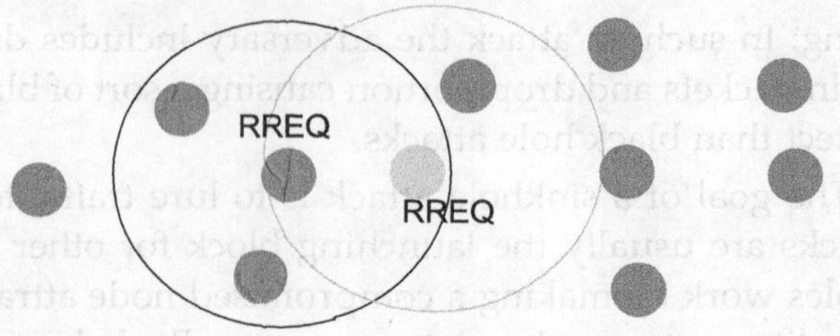

Fig. 11.5 Wireless sensor network routing dialogues

11.4.4 Transport Layer Attacks

The transport layer can be attacked via flooding or de-synchronization.

(*a*) The goal of flooding attacks is to exhaust memory resources of a victim system. Similar to TCP SYN attacks the attacker sends many connection establishment requests, forcing the victim to allocate memory in order to maintain the state for each connection.

(*b*) In de-synchronization attacks, flags and sequence numbers are usually modified. Attacker prevent the endpoints from ever exchanging messages as they will be continually requesting retransmission of previous erroneous messages that leads to an infinite cycle that wastes energy. This type of attack can pose a larger threat to the network.

Table 11.1 Sensor network layers and attacks/defenses

Network Layer	Attacks	Defenses
Physical	Jamming	Spread-spectrum, priority messages, lower duty cycle, region mapping, mode change
	Tampering	Tamper-proof, hiding
Link	Collision	Error correcting code
	Exhaustion	Rate limitation
	Unfairness	Small frames
Network and routing	Neglect and greed	Redundancy, probing
	Homing	Encryption
	Misdirection	Egress filtering, authorization monitoring
	Black holes	Authorization, monitoring, redundancy
Transport	Flooding	Client Puzzles
	Desynchronization	Authentication

In Table 11.1 shows sensor layers of a typical wireless sensor network that are summarized along with their attacks and defenses. One strategy is to identify the jammed part of the sensor network and effectively route around the unavailable portion against the classic jamming attack.

Finally, attacks launched on a network may be insider or outsider attacks. In outsider attacks the attacker has no special access to the network. In insider attacks however, the attacker is considered to be an authorized participant of the network. Such attacks are either launched from compromised wireless sensor nodes running malicious code from legitimate nodes.

11.5 INTRUSION DETECTION

The intrusion detection in wireless sensor networks is an important section that covers to detect attacks on the sensor network. Many secure routing schemes attempt to identify network intruders, and key establishment techniques are used in part to prevent intruders from overhearing network data.

Despite the necessity of effective intrusion detection schemes there is no good solution has devised for wireless sensor networks. Hence, several researches are on in the wireless sensor networks.

As such, it is difficult to define characteristics (or signatures) that are specific to a network intrusion as opposed to the normal network traffic, because this might result of normal network operations or malfunctions resulting from the environment change.

11.5.1 Intrusion Detection in Wireless Sensor Networks

Typically a wireless sensor network uses cryptography to secure itself against unauthorized external nodes gaining entry into the network. But, cryptography can only protect the network against the external nodes and does little to thwart malicious nodes that already possess one or more keys.

As per Brutch and Ko they classify intrusion detection systems (IDS) into following categories: host-based, network-based, signature based, anomaly based, and specification based.

- Host based IDS system operates on operating systems audit trails, system call audit trails, logs, and so on.
- Network based IDS, on the other hand, operates entirely on packets that have been captured from the network.
- Signature based IDS simply monitor the network for specific pre-determined signatures that are indicative of an intrusion.
- Anomaly based scheme, a standard behavior is defined and any deviation from that behavior triggers the intrusion detection system.
- Finally, Specification based scheme defines a set of constraints that are indicative of a program's or protocol's correct operation.

11.6 SECURITY REQUIREMENTS

A sensor network is a special type of network that shares wireless gadgets with a typical computer network. Therefore, the requirements of a wireless sensor network as encompassing both the typical and unique network requirements suited solely to wireless sensor networks.

11.6.1 Data Confidentiality

Confidentiality requirement is needed to ensure that sensitive information is well protected and not revealed to unauthorized third parties. In the sensors' environment the confidentiality protect information travelling between the sensor nodes of the network or between the sensors and the base station from disclosure,

Data confidentiality is the most important issue in network security that focus in sensor networks relates to the following:

- A sensor network should not leak sensor readings to its neighbors. Example especially in a military application, where, the data stored in the sensor node may be highly sensitive.
- In an application nodes communicate with highly sensitive data, e.g., key distribution; therefore it is extremely important to build a secure channel in a wireless sensor network.
- Public sensor information, such as sensor identities and public keys, should also be encrypted to some extent to protect against traffic analysis attacks.

The standard approach for keeping sensitive data secret is to encrypt the data with a secret key that only intended receivers possess, thus achieving confidentiality.

11.6.2 Data Integrity

Data integrity is about the safety of data. The adversary can change the data, to send the sensor network into disarray. For example, a malicious node may add some fragments or manipulate the data within a packet. This new packet can then be sent to the original receiver. So, the protection of such data loss and data integrity to ensure that any received data has not been altered in transit.

11.6.3 Data Freshness

Data freshness suggests that the data is recent, and it ensures that no old messages have been replayed. This is especially important for shared-key strategies to be propagated on the entire network. It is to solve the problem, identification such as a nonce, or time-related counter, that can be added into the packet to ensure data freshness.

11.6.4 Availability

Availability ensures that services and information can be accessed at the time when required. In sensor networks, there are many risks that could result in loss of availability such as sensor node capturing and denial of service attacks. Some approaches choose to modify the code to reuse as much also try to make use of additional communication to achieve the same goal. Even some approaches force strict limitations on the data access, or propose an unsuitable scheme to simplify the algorithm. But all these approaches weaken the availability of a sensor network for the following reasons:

- Additional computation consumes additional energy so if no more energy exists, the data will no longer be available.
- Additional communication also consumes more energy as per the increase in communication and it does a communication conflict.
- A single point failure will be introduced if using the central point scheme. This greatly overloads the availability of the network.

The requirement of security not only affects the operation of the network, but also is highly important in maintaining the availability of the whole network.

11.6.5 Self-Organization

A wireless sensor network is a typically an ad hoc network, that requires every sensor node to be independent and flexible enough to be self-organizing and self-healing according to different situations.

The distributed sensor networks must self-organize to support multihop routing, they must also self-organize to conduct key management and building trust relation among sensors. If self-organization it support multihop routing and key management that is lacking in a sensor network, the damage resulting from an attack or even the hazardous environment.

11.6.6 Time Synchronization

Most sensor network applications rely on some form of time synchronization. In order to conserve power, an individual sensor's radio may be turned off for periods of time. Furthermore, sensors may compute the end-to-end delay of a packet as it travels between two pair wise sensors. A more collaborative sensor network may require group synchronization for tracking applications.

11.6.7 Secure Localization

The effectiveness of a sensor network will rely on its ability to accurately and automatically locate each sensor in the network. Unfortunately, an attacker can easily manipulate non secured location information by reporting false signal strengths, replaying signals, etc.

A sensor computes its location by listening for the beacon information sent by each locator. A device's position is accurately compute the authenticated ranging and distance bounding that are used to ensure accurate location of a node.

11.6.8 Authentication

In the sensor network authentication is necessary to verify the data sent by the claimed sender. Authentication objective is to protect from false when sending data or either compromising exchanged data.

11.6.9 Freshness

Data freshness objective ensures that messages are fresh, meaning that they obey in a message ordering and have not been reused. One of the many attacks launched against sensor networks is the message replay attack where an adversary may capture messages exchanged between nodes and replay them later to cause confusion to the network. To achieve freshness, network protocols must be designed in a way to identify duplicate packets and discard them preventing potential mix-up.

11.6.10 Secure Management

Management is required in every system that is constituted from multi components and handles sensitive information. In the case of sensor networks, secure management is on base station level; since issues like key distribution to sensor nodes to establish encryption and routing information that need secure management. Therefore, secure protocols for group management are required for adding and removing members, and authenticating data from groups of nodes.

11.7 CHALLENGES

The severe constraints and demanding deployment environments of wireless sensor networks make computer security for the systems more challenging than for conventional networks. However, several properties of sensor networks may help address the challenge of building secure networks. Some of important challenges are:

- To develop an architect security solution for the systems in their early design and research stages.
- To develop applications that is likely to involve the deployment of sensor networks under a single administrative domain, simplifying the threat model.
- To exploit redundancy, scale, and the physical characteristics of the environment and to resist from further attack. Many other problems also need further research.
- To design protocol for secure wireless communication links against eavesdropping, tampering, traffic analysis, and denial of service.
- To develop asymmetric public-key cryptosystems protocols for low-end devices on the base station efficiently.
- Finally, finding ways to tolerate the lack of physical security, perhaps through redundancy or knowledge about the physical environment, will remain a continuing overall challenge.

11.7.1 Research Challenges

The severe constraints and demanding deployment environments of wireless sensor networks make computer security for these network systems more challenging than for conventional networks. However, several properties of sensor networks address the challenge of building secure networks. First, the development on the architect of security solutions, second, many applications are likely to involve the deployment of sensor networks under a single administrative domain, simplifying the threat model. Third, it may be possible to exploit redundancy, scale, and the physical characteristics of the environment in the solutions.

Ultimately, the unique aspects of sensor networks allow novel defenses not available in conventional networks. Many other problems also need further research is to secure wireless communication links against eavesdropping, tampering, traffic analysis, and denial of service. Others involve resource constraints. Ongoing directions include asymmetric protocols where most of the computational burden falls on the base station and on public-key cryptosystems efficient on low-end devices. Finally, finding ways to tolerate the lack of physical security, perhaps through redundancy or knowledge about the physical environment, will remain a continuing overall challenge.

SUMMARY

In this Chapter, discussed the security issues in wireless sensor networks at different layers and conditions. Some different prone areas security aspects discussed. In wireless sensor network security: obstacles, requirements, attacks, and defenses which can leads to issues like routing, trust, denial of service, and so on. Hence, it is important to have broad area of wireless sensor network security that can be interested for the researcher.

As wireless sensor networks continue to grow and become more common, the further expectations of security will be required for these wireless sensor network applications. There is an expectation that the current and future work in privacy and trust will make wireless sensor networks a more attractive option in a variety of new arenas.

There are many failure nodes in wireless sensor networks disposed in hostile and unattended environments. These failure nodes are caused by random fault, malice intrusion, or exhausted battery. The goal is to achieve a topology with high node-failure tolerance with the existence of intrusion nodes. The secure topology control method and the security algorithm make the networks to sustain the node-failure tolerant ability and will prolong network battery life.

QUESTIONS

1. What are the impediments of wireless sensor network security?
2. Discuss the threats in layers of WSN in detail.
3. Brief out the Intrusion Detection in Wireless Sensor Networks.
4. Why are the security requirements in wireless sensor networks?
5. What are the challenges in the wireless sensor networks?

BIBLIOGRAPHY

- Wireless Sensor Network Security: A Survey, John Paul Walters, Zhengqiang Liang, Weisong Shi, and Vipin Chaudhary, 2006 Auerbach Publications, CRC Press.
- R. Anderson and M. Kuhn. Low cost attacks on tamper resistant devices. In IWSP: International Workshop on Security Protocols, LNCS, 1997.
- H. Chan and A. Perrig. Security and privacy in sensor networks. IEEE Computer Magazine, pages 103-105, 2003.
- J. Deng, R. Han, and S. Mishra. Security, privacy, and fault tolerance in wireless sensor networks. Artech House, August 2005.
- Wood, A. and Stankovic, J. Denial of service in sensor networks. IEEE Comput. (Oct. 2002), 54-62.
- Wireless Sensor Network for Security: Issues and challenges, Tolga Onel, Ertan Onur, Cem Ersoy and Hakan Delic.
- Security Topology Control Method for Wireless Sensor Networks with Node-Failure Tolerance Based on Self-Regeneration, Liang-MinWang, Yuan-Bo Guo, and Yong-Zhao Zhan, EURASIP Journal on Wireless Communications and Networking, Volume 2010.
- Perrig, R. Szewczyk, V. Wen, D. Culler, and J. D. Tygar, "SPINS: security protocols for sensor networks," in Proceedings of the 7th Annual International Conference on Mobile Computing and Networking (MOBICOM '01), pp. 189-199, ACM Press, Rome, Italy, July 2001.
- Security Models for Wireless Sensor Networks, Sophia Kaplantzis, March 20, 2006.
- Karlof and D. Wagner, "Secure routing in wireless sensor networks: Attacks and countermeasures," Ad-Hoc Networks, vol. 1, no. 2-3, pp. 293-315, 2003.
- E. Loo, M. Y. Ng, C. Leckie, and M. Palaniswami, "Intrusion detection for sensor networks," 2004.

12

ENERGY HARVESTING IN WSN

12.1 INTRODUCTION

In recent years, Wireless Sensor Networks (WSNs) are an exciting new area of research. Technological developments over the last decade have yielded several specialized embedded platforms capable of sensing, computing and communication. One of the very scarce resources for these types of networks is energy. These networks are expected to have a long lifetime (weeks to years) without human intervention for energy replenishment. Human intervention is undesirable since large number of nodes imply high operational cost.

Energy harvesting exploits the technologies to generate electricity from the environment, which can be used to power electronics and electrics. Different types of technologies can be employed depending on the energy source. In the wireless sensor networks (WSNs) research has predominantly use of a portable and limited energy source, viz. batteries, in the power sensors. There is a negative feature in this technology. Without energy, a sensor is essentially useless and cannot contribute to the effectiveness of the network as a whole. Consequently, substantial research efforts have been carried on designing energy-efficient networking protocols to maximize the lifetime of WSNs. However, the emerging WSN applications where sensors are required to operate for much longer durations should have a battery capacity to withstand.

Energy harvesting has reached an incline point because the necessary lower power electronics and more efficient energy gathering and storage are now sufficiently affordable, reliable and longer lived for a huge number of applications to be practicable. Some of such examples are in environmental/habitat monitoring and structural health monitoring of critical infrastructures and buildings, where batteries are hard (or impossible) to replace/recharge. Also, the wireless sensor network in the areas of habitat and environmental monitoring to represent an enormous potential benefits for scientific communities and society as a whole are another dimension.

This evaluates the tradeoffs between different approaches to implementing several sensor network services. However, for all these harvesting the energy is important; which is to convert the ambient energy from the environment into electricity to power the sensor nodes. While renewable energy technology is not new (example solar and wind) the systems in use are far too large for

WSNs. Those small enough for use in wireless sensors are most likely able to provide only enough energy to power sensors eliminates the need of battery replacement.

A few instantiations of such deployments are for applications such as, Habitat monitoring - environmental monitoring in wild-life habitats like Great Duck Island and James Reserve, to study weather conditions and animal migratory patterns etc. As seen in applications mentioned above, battery-powered sensor nodes can be used to instantiate applications in remote locations where power lines do not exist. A tradeoff of these benefits is the finite battery capacity. Finite node lifetime implies finite lifetime of the applications or, additional cost and complexity to regularly change batteries. Nodes could possibly use large batteries for longer lifetimes, but will have to deal with increased size, weight and cost. Nodes may use low-power hardware like low-power processor and radio, at the cost of lesser computation ability and lower transmission ranges, respectively.

An alternative technique that has been applied to address the problem of finite node lifetime is the use of energy harvesting. Energy harvesting refers to harnessing energy from the environment or other energy sources and converting it to electrical energy. The harnessed electrical energy powers the sensor nodes. Further, as mentioned above, sensor nodes can exploit energy harvesting opportunities to dynamically tune system parameters.

12.2 RESEARCH ON ENERGY HARVESTING

Energy is everywhere in the environment surrounding us - available in the form of thermal energy, light (solar) energy, wind energy, and mechanical energy. The developments and demand have increased the efficiency of devices in capturing trace amounts of energy from the environment and transforming them into electrical energy. Figure 12.1 shows the Energy Harvesting Feasibility about future and the power demand growth rate due to high computation environment.

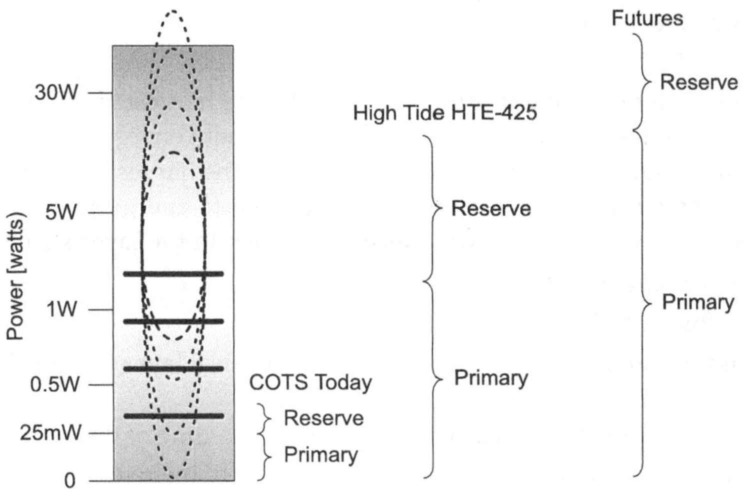

Fig. 12.1 Energy harvesting feasibility

Energy harvesting (EH) from a natural source, where a remote application is deployed, and where such natural energy source is essentially inexhaustible, is an increasingly attractive alternative to inconvenient wall plugs and costly batteries. This free energy source, is available maintenance-free and is now available throughout the lifetime of the application. In addition, energy harvesting is used as an alternative energy source to supplement a primary power source and to enhance the reliability of the overall system and to prevent power interruptions.

The Renewable energy that is being harvested to generate electricity today includes thermal, light, solar, wind, water, electromagnetic energy and thermal energy. Harvesting energy is a low-power devices like wireless sensors presents a new challenge as the energy harvesting device in the present power constraint issue for small scale sensor device. There are certain complex tradeoffs to be considered when designing energy harvesting circuits for WSNs arising due to various factors like the energy sources, energy storage device, power management functionality of the nodes and protocols, and the applications' requirements.

12.3 ARCHITECTURE OF ENERGY HARVESTING SENSOR NODES

Energy harvesting refers to scavenging energy or converting energy from one form to the other source. It is applied to sensor nodes, energy from external sources can be harvested to power the nodes and in turn, increase their lifetime and capability. Given the energy-usage profile of a node, energy harvesting techniques build up the demand of energy needs. A widespread and popular technique of energy harvesting is converting solar energy to electrical energy.

Solar energy is uncontrollable—the intensity of direct sunlight cannot be controlled—but it is a predictable energy source with daily and seasonal patterns. Other example of energy harvesting is to convert mechanical energy or wind energy in to electrical energy. For example, mechanical stress applied to piezo-electric materials, or to a rotating arm connected to a generator, can produce electrical energy.

A typical energy harvesting system has three components,

I. Energy source,
II. Harvesting architecture and
III. Load.

Energy source refers to the ambient source of energy to be harvested. Harvesting architecture consists of mechanisms to harness and convert the input ambient energy to electrical energy. Load refers to activity that consumes energy and acts as a sink for the harvested energy. The low power architecture and low power network design is the approach to energy management at different communication layers. Low power network design approach in the layer structure follows as:

I. Low power hardware architectures
II. Low power software techniques
III. Limiting transmission range and power control at physical layer to bound device consumption.
IV. Low power MAC mainly by increasing MAC layer sleep time of the nodes.
V. Dynamic configuration of nodes with extra deployment of them in any geographic region for sleep cycles in higher time granularity.
VI. Geographic and power aware routing to bound network traffic.
VII. Data Aggregation to increase the good put of the network and to suppress unnecessarily data traffic.

Wireless networked sensors and actuators deeply embedded in the day-to-day life. The Fig. 12.2 shows the energy harvesting using wireless sensor network consist of sensor motes placed at various location as per the application and it is energized by the various harvesting sources. Each sensor nodes collect the information and process through the gateway to the internet network.

Various Individual sensor nodes communicate and coordinate with one another and take several forms. To provide the data to remote end-users location, the base station includes WAN connectivity and persistent data storage for the collection of sensor patches. The architecture of the wireless sensor network includes sensor nodes, gateways, base stations that has some persistent storage and also provides data management services.

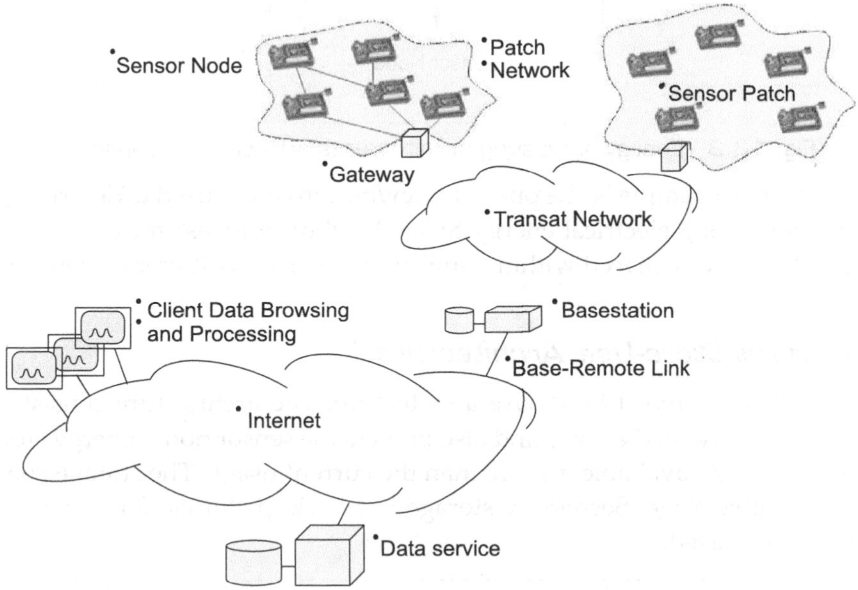

Fig. 12.2 Architecture of harvesting wireless sensor network

A possible solution to this crisis is to harvest the energy from the physical environment at run time. The various source of energy that is harvested for the sensor node. Here the most common sources for ambient energy are light, thermal, RF and vibration. Every harvesting has unique advantages and disadvantages, and the specific harvesting technology is dependent on the application.

12.3.1 Energy Harvesting Architecture

Habitat study is one of the driving applications for WSNs. Such applications usually require the sensing and gathering of bio-physical or bio-chemical information from the entities under study, such as Redwoods, Storm Petrels, Zebras, and Oysters. In many scenarios, habitat study requires relatively simple signal processing, such as data aggregation using minimum, maximum, or average operations.

Broadly, energy harvesting can be divided into two architectures:

I. Harvest-Use : Energy is harvested just-in-time for use and
II. Harvest-Store-Use: Energy is harvested whenever possible and stored for future use.

12.3.1.1 Harvest-Use Architecture

In the case of Fig. 12.3 which is a Harvest-Use architecture, the harvesting system is directly powers the sensor node and as a result, for the node to be operational, the power output of the harvesting system has to be continuously above the minimum operating point. The energy variations in har-vesting capacity close to the minimum power point will cause the sensor node to oscillate in ON and OFF states.

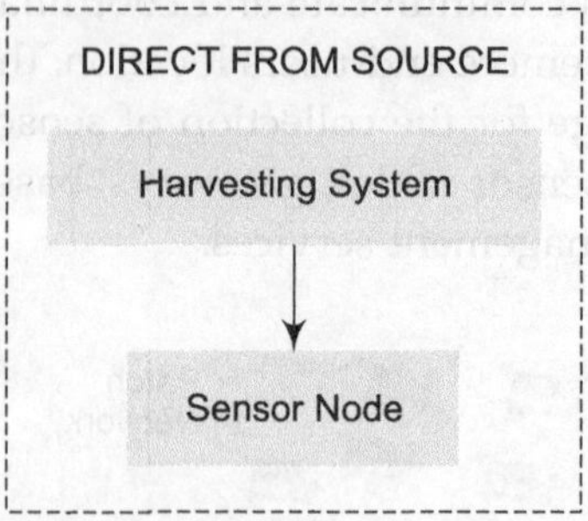

Fig. 12.3 Energy harvesting architectures with storage capability

A Harvest-Use system example is, the push of a key/button can be used to deform a piezo-electric material, thereby generating electrical energy to send a short wireless message. Similarly, piezo-electric materials strategically placed within a shoe may deform to different extents while walking and running.

12.3.1.2 Harvest-Store-Use Architecture

Figure 12.4 portrays the Harvest-Store-Use architecture. The architecture consists of a storage component that stores harvested energy and also powers the sensor node. Energy storage is useful when the harvested energy available is more than the current usage. The storage component may be single-stage or double-stage. Secondary storage is a backup storage for situations, where the Primary storage is exhausted.

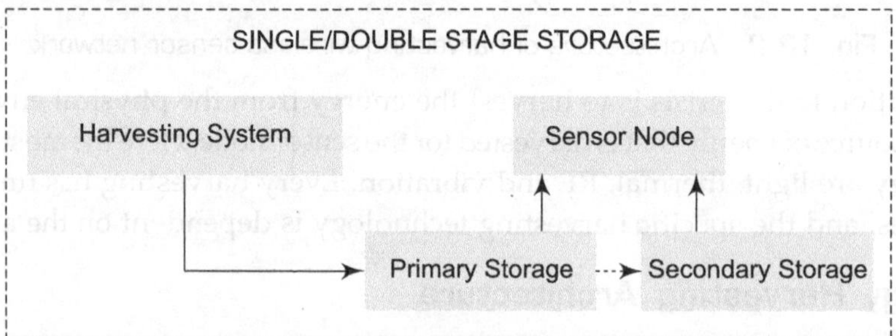

(b) Harvest-Store-Use

Fig. 12.4 Energy harvesting architectures without storage capability

As an example, a Harvest-Store-Use system can use uncontrolled but predictable energy sources like solar energy. During the daytime, energy is used for work and also stored for later use. During night, the stored energy is conservatively used to power the sensor node.

12.4 HARVESTING ENERGY SOURCE

Energy harvesting sources have different characteristics of controllability, predictability and mag-nitude. A controllable energy source can provide harvestable energy whenever required; energy availability need not be predicted before harvesting. With non-controllable energy sources, either energy may be simply harvested whenever available or, harvesting opportunities may be sched-uled. Further, energy sources can be broadly classified into the following two categories,

I. Ambient Energy Sources : Sources of energy from the surrounding environment, e.g., solar energy, wind energy and RF energy, and

II. Human Power: Human power refers to the energy harvested from body movements of humans.

Passive human power sources are those which are not user controllable. Some examples are blood pressure, body heat and breath. Active human power sources are those that are under user control, and the user exerts a specific force to generate the energy for harvesting, e.g., finger motion, paddling and walking.

Table 12.1 tabulates characteristics of different energy sources as fully controllable, partially controllable, uncontrollable, but predictable and uncontrollable and unpredictable.

Table 12.1 Listing and characterization of energy sources

Energy Source	Characteristics
Solar	Ambient, Uncontrollable, Predictable
Wind	Ambient, Uncontrollable, Predictable
RF Energy	Ambient, Partially controllable
Body heat, Exhalation, Breathing and Blood Pressure	Passive human power, Uncontrollable, Unpredictable
Finger motion and Footfalls	Active human power, Fully controllable
Vibrations in indoor environments	Ambient, Uncontrollable, Unpredictable

12.4.1 Common Sources of Energy Harvesting

I. Mechanical Energy – from sources such as vibration, mechanical stress and strain

II. Thermal Energy – waste energy from furnaces, heaters, and friction sources

III. Light Energy – captured from sunlight or room light via photo sensors, photo diodes, or solar panels

IV. Electromagnetic Energy – from inductors, coils and transformers

V. Natural Energy – from the environment such as wind, water flow, ocean currents, and solar

VI. Human Body – a combination of mechanical and thermal energy naturally generated from bio-organisms or through actions such as walking and sitting

VII. Other Energy – from chemical and biological sources

12.4.2 Harvesting Energy

The Energy harvesting is defined as the conversion of ambient energy into usable electrical energy. Energy harvesting is a technology capture unused ambient energy (such as vibration, strain, temperature gradients, energy of gas and liquid flows) and convert into usable electrical energy that is stored and used for sensing or actuation. The environment represents a relatively inexhaustible source of energy when compared with the energy stored in common storage elements, like batteries.

Energy harvesting is a perfect match for wireless devices and wireless sensor networks that otherwise relies on battery power. Consequently, energy harvesting (*i.e.*, scavenging) methods must be characterized by their power density, rather than energy density. Below describes the different source of harvesting energy.

12.4.3 Solar Energy

Photovoltaic cells convert incident light into electrical energy. In solar energy based energy is based on the principle of photo-voltaic effect. The recharging circuitry is designed, to recharges the batteries of the nodes when the charge drops below a threshold level. This may leads to strengthening the lifetime of the nodes as well as the network. Here the compatibility and low power design are the two major salient features of the designed circuitry. The performance of the circuit is also tested with the MicaZ hardware.

12.4.4 Mechanical

The vibrations are present in all systems. The vibration, kinetic and mechanical energy generated the harvesting vibration energy. The PMG7 micro generator is the vibration energy harvester to power wireless and battery-free devices capable of sending large amounts of data from many types of industrial equipment.

12.4.5 Thermal

Thermal gradients in the environment are directly converted to electrical energy through the thermoelectric effect. The generated voltage and power is proportional to the temperature. Large thermal gradients are essential to produce practical voltage and power levels.

12.4.6 Electromagnetic Energy

An electromagnetic harvester uses the principle of a spring-mass system to convert energy of vibration into electrical energy. The term electromagnetic field or radiofrequency (RF) field may be used to indicate the presence of electromagnetic or RF energy. Where the RF energy transceiver system of wireless sensor with power requirements in the <100mW range with broadcast RF energy, the receiver can pick up and convert any RF energy regardless of frequency. This is in turn charge the battery devices.

Some other energy which is a renewable/non renewable source of energy also contributes towards the part of energy harvesting.

Table 12.2 compares the estimated power and challenges of some important ambient energy sources. For instance, Light, can be a significant source of energy, but it is highly dependent on the application and the exposure to which the device is subjected. On the other hand, Thermal energy is limited, because the temperature differentials across a chip are typically low. Vibration energy is a moderate source, but again dependent on the particular application and the Electromagnetic wave result in RF energy which is present everywhere can be accessed in remote location also.

Table 12.2 Comparison between different ambient energy sources

Energy Source	Efficiency	Estimated Power
Light	10-25%	10μW-15mW (Outdoors: 0.15mW-15mW) (Indoors: <10μW)
Vibrations	25-50%	1μW-200μW (Piezoelectric: ~ 200μW) (Electrostatic:50μW-100μW) (Electromagnetic: <1μW)
Thermal	-0.1-3%	15μW (10°C gradient)
Electromagnetic	50%	10μW-100μW (2.4 GHz)

12.5 ENERGY HARVESTING SENSOR NODES USING SOLAR PANEL

The implementations of energy harvesting sensor nodes designed to use various energy sources such as solar energy, active user power, wind energy and RF energy. These solar energy harvesting implementations are different along axes like characteristics of solar panels, battery type, capacity, and recharge circuit.

Fig. 12.5 Photo of heliomote prototype architecture

The Fig. 12.5 shows a picture of the Heliomote prototype and its system architecture. The Heliomote a single-storage energy harvesting system for scavenging solar energy built using the Mica2 platform. Heliomote uses a solar panel of area 3.75 inches x 2.5 inches which outputs 60mA at a voltage of 3.3V.

The power from this solar panel is used to recharge two AA-sized Ni-MH battery of capacity 1800mAh each. Overcharging a Ni-MH battery can lead to instability and is very hazardous, and so an overcharge protection circuit is used. Similarly, undercharging the Ni-MH battery is prevented so that the load does not continue drawing power even after the battery voltage has dropped below a low threshold.

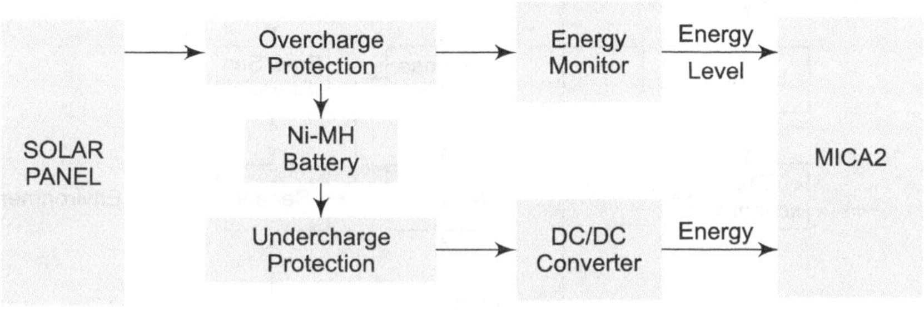

Fig. 12.6 Photo of the heliomote architecture

The Fig. 12.6 shows a picture of Heliomote, both the over-charge and under-charge protection modules are hardware components (control circuits) consisting of comparator with hysteresis, that control analog switches. Heliomote also has an energy monitoring component which enables a sensor node to learn its energy availability and usage. The Energy Monitor component of the Heliomote measures and conveys information regarding the amount and variance of energy extracted.

Using energy harvesting to supplement batteries does not eliminate the problem of having to replace the batteries when they run out. The process merely delays the inevitable. In applications like structural health monitoring of civil infrastructures, the sensors need to be installed and operate for long durations, from years to decades or even longer. Combining low-power electronics, energy harvesting devices, and super capacitors, it is possible to implement harvesting wireless sensor.

The state-of-the-art energy harvesting devices is in the MICAz based mote, the results are shown in Table 12.3.

Table 12.3 Result of Duty Cycle by MICAz with 10cm^2 Harvesting material

Technology	Power Density [29] (μW/cm^2)	Energy Harvesting Rate (mW)	Duty Cycle (%)
Vibration – electromagnetic	4.0	0.04	0.05
Vibration – piezoelectric	500	5	6
Vibration – electrostatic	3.8	0.038	0.05
Thermoelectric	60	0.6	0.72
Solar – direct sunlight	3700	37	45
Solar – indoor	3.2	0.032	0.04

12.6 DESIGN OF ENERGY HARVESTING WSN

The design of Harvesting wireless sensor network is shown in Fig. 12.7 for the usual implementation for the researchers.

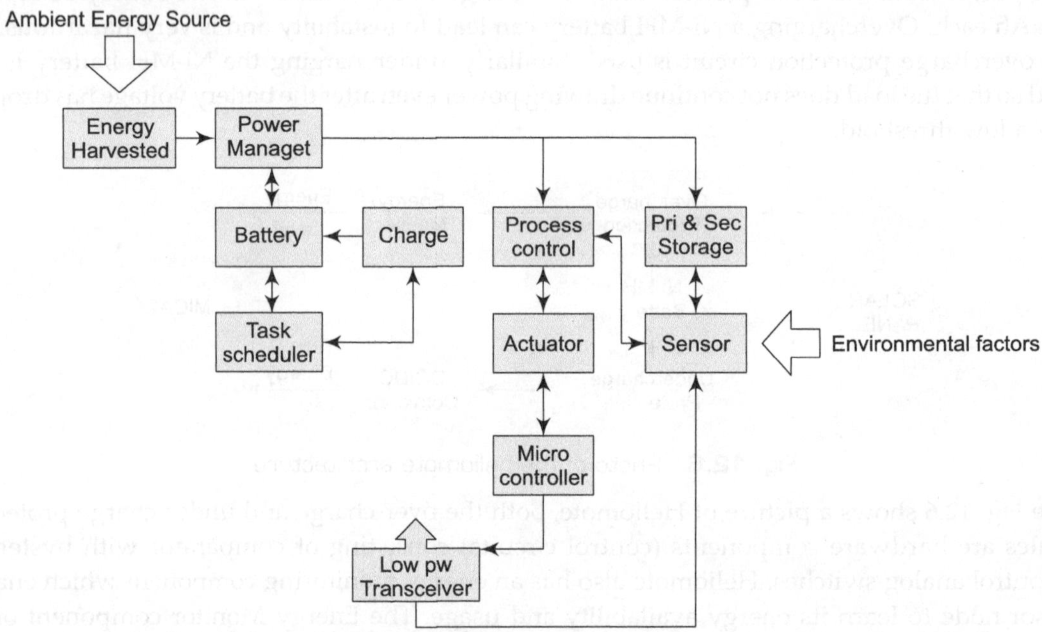

Fig. 12.7 Design of harvesting wireless sensor network

As per the figure the energy harvesting device receives the harvesting energy from the ambient environment and as per the application that the sensor node actuates the conversion of ambient energy to electrical energy occurs. This energy follows the characteristics that a sensor network to monitor the power management and the battery charging.

In the example of solar panel as shown in Fig. 12.8, the power from this energy harvesting sensor used to recharge two AA-sized Ni-MH battery of capacity 1800 mAh each. The solar panel of area 3.75 inches by 2.5 inches outputs 60mA at a voltage of 3.3V for charging the battery.

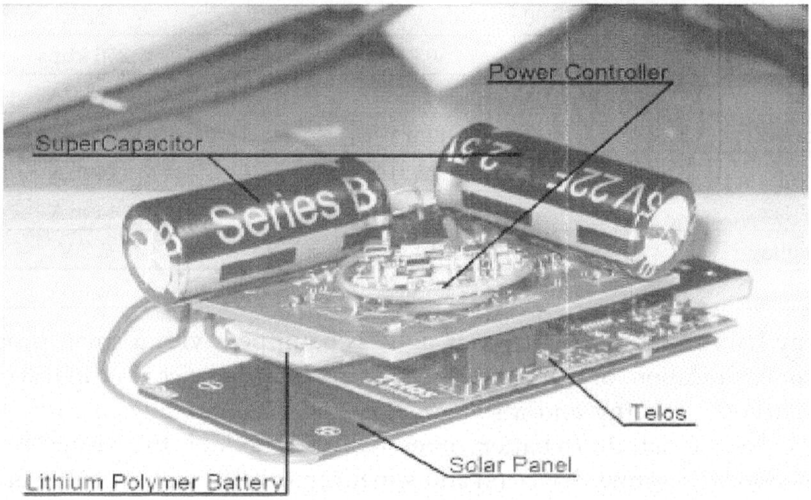

Fig. 12.8 Solar panel energy harvesting.

Without the sustained energy supply, the exact sleep and wakeup timings are unknown. Therefore, the operating characteristic of Energy Harvesting-WSN as compared to battery-operated WSN is shown in Fig. 12.9.

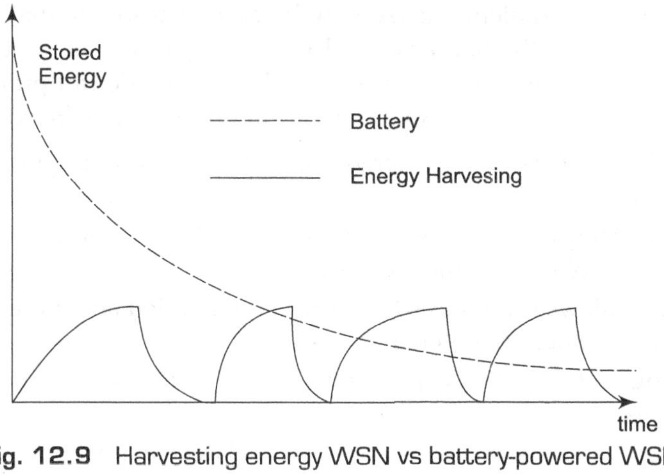

Fig. 12.9 Harvesting energy WSN vs battery-powered WSN

The over-charge and under-charge protection modules are hardware components (control circuits) consisting of comparator with hysteresis, that control analog switches. Table 12.4: summarize of power consumption of commercial sensor network nodes.

Table 12.4 Summary of power consumption of commercial sensor network nodes

	Crossbow- MICAz	Intel- Mote2
Radio standard	IEEE 802.15.4/ZigBee	IEEE 802.15.4
Typical range	100m outdoor, 30m indoor	30m
Data rate (kbps)	250 kbps	250 kbps
Sleep mode	15 μA	390 μA
Processor only	8 mA	31-53 mA
RX	19.7 mA	44 mA
TX	17.4 mA	44 mA
Supply voltage	2.7 V	3.2 V
Average	2.8 mW	12 mW

In WSN energy harvesting of, each sensor node is equipped with a small processor, a radio transceiver for communication, one or more energy harvesting devices, a capacitor/super capacitor to store the harvested energy and a sensor. The key differentiating feature of a Harvesting Energy WSN node (as compared to a battery-operated WSN node) is the energy source which is a combination of energy harvesting device (s) and super capacitor(s) instead of batteries. The sink is powered by an external source and remains on all the time.

12.7 BENEFIT OF ENERGY HARVESTING

Energy harvesting provides numerous benefits to the end user. Its solutions can reduce dependency on battery power. Harvested ambient energy may be sufficient to eliminate battery completely. The device can be powered by the harvester and rely on internal energy storage to smooth out variations in available ambient energy. For example, thermoelectric can produce watts per cubic centimeter, while piezoelectric in actuators and vibration harvesters can exhibit 60% efficiency.

I. It reduces installation costs. Self-powered wireless sensors do require wires, conduits and are very easy to install.

II. As much as 38% of energy is consumed in buildings, so long life and more affordable building controls are the focus of most of the companies.

III. Energy harvesting allows for devices to function unattended and eliminates maintenance to replace batteries. It reduces maintenance costs.

IV. It provide sensing and actuation capabilities in hard-to-access hazardous environments on a continuous basis.

V. Provide long-term solutions. A reliable self-powered device functions virtually as long as the ambient energy is available. The self-powered devices are perfectly suited for long-term applications.

VI. It reduces environmental impact. Energy harvesting can eliminate the need for large number of new batteries and energy costs of battery replacements.

VII. The life of energy-harvesting devices without batteries is at least 10 years longer than the life of battery-driven wireless devices and this gives valuable paybacks.

VIII. The development of next generation of energy-harvesting devices will be smaller, lower in cost and its lives will exceed 20 years in many cases.

The energy density will improve research scope for the efficiency and that will expand the addressable market. Many devices will be announced that work within the human body without need of further intrusive surgery and many other applications. Many others will be embedded in apparel. Energy harvesting will power many of the billions of medical disposables needed for self-diagnostics and drug delivery in the future days.

12.8 CHALLENGES OF ENERGY HARVESTING WSN

As the wireless sensor network growth is rising in tremendous pace the challenge, the ad-hoc wire-less network is soaring. The routing protocol that is harvesting-aware is higher in cost, or whose nodes have depleted batteries. While there has been extensive research on wireless sensor networks, those specific to energy harvesting WSNs are just emerging. Some of such issues are in the areas such as:

I. Topology control
II. Deployment issue
III. MAC issue
IV. Power Management
V. Network & protocol issue
VI. Reliable data release
VII. Data Delivery Schemes
VIII. Routing algorithm
IX. Design factors
X. Energy Storage Technology
XI. Energy harvesting for infinite life time sensor nodes
XII. Security

Also, due to distributed processing capability the issue in sensor network has a greated impact on battery. Because each sensor node is operated by battery with finite energy storage and the lifetime of nodes depends on the available energy in the battery. Hence, these issues highlighted above need to address.

SUMMARY

Wireless sensor networks that are powered by ambient energy harvesting are a promising technology for many sensing applications as it eliminates the need to replace batteries. In this chapter, discussed the current state of technology in energy harvesting is discussed a sustained energy supply to enable WSNs continuously.

The chapter focuses on harvesting energy. In the wireless sensor networks, the use of renewable and non-renewable energy source is a promising technology for harvesting many sensing applications as it eliminates the need to replace batteries. Levels of harvested energy may vary significantly from application to application.

Furthermore, the ability to harvest energy from the environment is highly dependent on many environmental factors and there need further research to understand and exploit. Here, in this chapter discussed the various issues related to energy harvesting and a method for doing still further in these research areas.

QUESTIONS

1. Why the energy harvesting is important study in WSNs?
2. Explain the architecture of energy harvesting sensor nodes.
3. Write note on Harvest-Store-Use Architecture.
4. What are the different sources of harvesting energy?
5. Explain the design block of Harvesting wireless sensor network.
6. What are the benefits of harvesting energy?

BIBLIOGRAPHY

- A.P. Chandrakasan et al., "Design Considerations for Distributed Microsensor Systems," Proc. Custom Integrated Circuits Conf., IEEE, Piscataway, NJ, 1999, pp. 279-286.
- Towards unified radio power management for wireless sensor networks, Guoliang Xing, Mo Sha, Greg Hackmann, Kevin Klues, Octav Chipara and Chenyang Lu Wireless Communication and Mobile Computing, (2008).
- Sentry-Based Power Management in Wireless Sensor Networks, Jonathan Hui, Zhiyuan Ren, and Bruce H. Krogh, _c Springer-Verlag Berlin Heidelberg 2003.
- Wireless Sensor Networks: A Survey on Ultra-Low Power-Aware Design, Itziar Marín, Eduardo Arceredillo, Aitzol Zuloaga, and Jagoba Arias.
- Experimental Measurements of the Power Consumption for Wireless Sensor Networks, Javier Alonso, Sergio Gómez, Miguel Alejandrez, Marisa Gil, Nacho Navarro.
- Energy Harvesting Sensor Nodes: Survey and Implications, Sujesha Sudevalayam Purushottam Kulkarni, December 19, 2008.
- C. Alippi, G. Anastasi, C. Galperti, F. Mancini, M. Roveri, "Adaptive Sampling for Energy Conservation in Wireless Sensor Networks for Snow Monitoring Applications", Proc. of IEEE International Workshop on Mobile Ad-hoc and Sensor Systems for Global and Homeland Security (MASS-GHS 2007), Pisa (Italy), October 8, 2007.
- S. R. Gandham, M. Dawande, R. Prakash, S. Venkatesan, "Energy efficient schemes for wireless sensor networks with multiple mobile base stations", Proc. of IEEE Globecom 2003, vol. 1, pp. 377-381, San Francisco, CA (December 1-5, 2003).
- IEEE Pervasive Computing, "Energy Harvesting and Conservation", Vol. 4, Issue 1, Jan-Mar. 2005.
- Energy Harvesting Sensor Nodes: Survey and Implications, Sujesha Sudevalayam Purushottam Kulkarni, December 19, 2008.
- Churchill, D.L., Hamel, M.J., Townsend, C.P., Arms, S.W., "Strain Energy Harvesting for Wireless Sensor Networks," Proc. SPIE's 10th Int'l Symposium on Smart Structures & Materials, San Diego, CA. Paper presented March, 2003.
- A. Mainwaring, D. Culler, J. Polastre, R. Szewczyk, and J. Anderson. Wireless sensor networks for habitat monitoring. In WSNA '02: Proceedings of the 1st ACM international workshop on Wireless sensor networks and applications, pages 88{97, New York, NY, USA, 2002. ACM Press.

13 Applications in WSN

13.1 INTRODUCTION

Wireless Sensor Network (WSN) has come to the forefront of the scientific community recently. This has been heralded as one of the most important technologies for 21st century. In recent years, the unique features of wirelessly operated sensor devices propelled researchers to identify potential applications for these relatively novel networks. Today, sensor networks hold the promise of improving processes and conditions in many areas as well as leading to entirely unforeseen opportunities. Today the developing countries, also growing demand into the applicability of wireless sensor networks brought prevailing challenges and actual needs.

WSN has been applied in many engineering fields, ranging from national defense and military affairs, to behavior observation of animals, structural health monitoring, traffic controls, medical treatment and sanitation and disaster monitoring. The applications of sensor networks are glimpsed here in this chapter.

13.2 APPLICATION IN WSN

A wireless sensor network can only be helpful if there is a substantial need or applications. The applications of wireless sensor networks are in the following application areas:

- Environmental Observation and Forecasting
- Disaster Prevention
- Agricultural Management
- Structure Health Monitoring
- Habitat Monitoring

Concrete applications that can be assigned to the first three application areas more or less directly affect the people's living conditions. An earthquake or volcano eruption warning system and monitoring of hazardous zones on a production plant can increase safety and prevent devastating incidents. Similarly, the ability to retrieve soil moisture in real time enables efficient irrigation and agricultural planning which is especially important in semi-arid regions of developing countries. The same holds for habitat monitoring, where wildlife can be studied without unnecessary human intrusion in remote areas. The application is explained below.

13.2.1 Environmental Observation and Forecasting

In the area of environmental observation and forecasting for warning system to protect the population, and on the other hand, provide researchers to study certain phenomena. This is because instrument in natural places, such as national parks, volcano's, riverbanks, rift zones, and woods with numerous networked sensor nodes can enable long-term data collection by each sensor node to provide localized measurements and detailed information. These features make the following applications feasible and beneficial in the context of developing countries:

- Volcanic Studies and Eruption Warning System
- Meteorological Observation
- Fire Detection
- Earthquake Studies and Warning System
- Water Quality Monitoring
- Flood, Cyclone and Tsunami Warning System

Dependent on the geographic region, one or several of these applications are conceivable. A good warning system can effectively help to mitigate the damages caused by natural disasters. Hence, the development of wireless sensor networks to assist meteorologists, geologists, and volcanologists has a great deal of importance in many less developed parts of the world. Nowadays sensor networks are also widely applied in habitat monitoring, agriculture research, fire detection and traffic control.

In the case of Bush Fire Response a low cost distributed sensor network is used for environmental monitor and disaster response. An integrated network of sensors combining on the ground sensors monitoring local moisture levels, humidity, wind speed and direction, together with satellite imagery and longer term meteorological forecasting that determine fire risk levels in targeted regions on probable fire direction. Such a network will provide valuable understanding of bushfire development and most importantly assist authorities in organizing a coordinated disaster response that saves lives and property by providing early warning for high risk areas.

Fancy Californian Winemaking is a project from Intel to find the right nutrients, exactly the right watering. The Fig. 13.1 shows the project made from Berkeley motes are installed in the test site - an Oregon, USA vineyard.

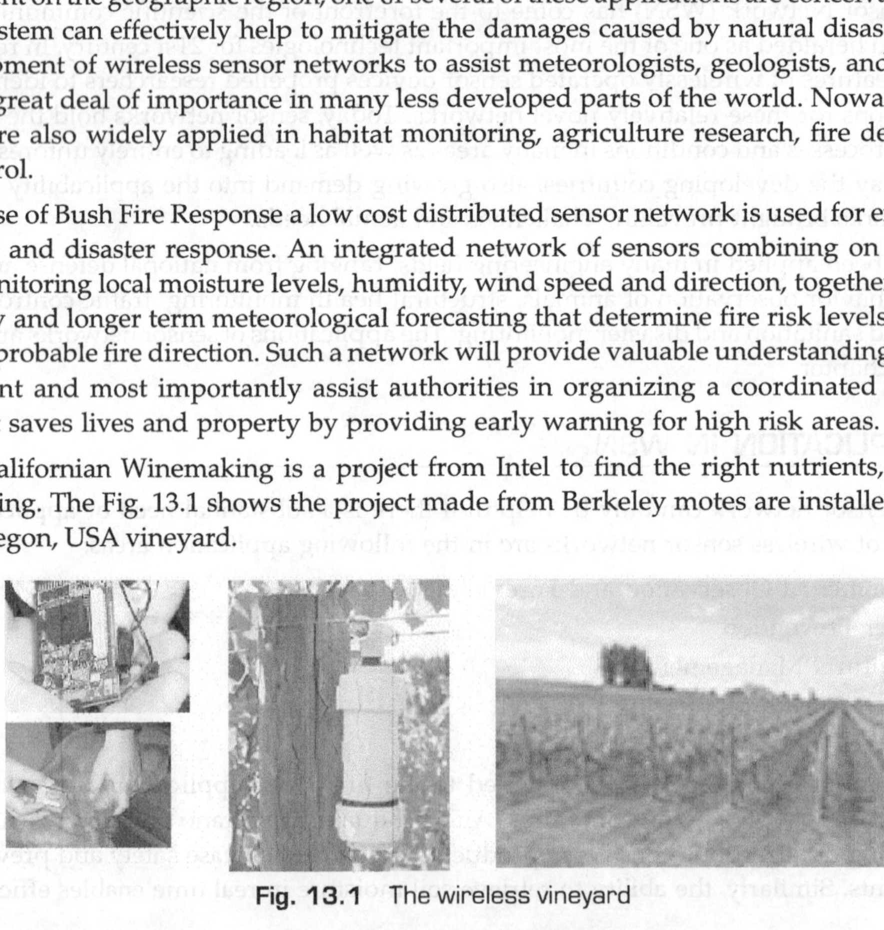

Fig. 13.1 The wireless vineyard

The sensor monitors temperature throughout the vineyard, where the temperature reading per minute and stores the results. The mote records the highest and lowest temperature readings for each hour of the day.

This sensor could monitor soil moisture and gather information to guide irrigation or harvesting to improve quality and to maximize yields. The further aim of this project with the help of sensor networks the owner of vineyard can manage the vineyard works more efficiently and automatically.

13.2.2 Military Applications

Wireless Sensor Network is a low power and small size sensor nodes. The sensor networks are applied very successfully in the military sensing. Now, wireless sensor networks application become an integral part of military command, control, communications, computing, intelligence, surveillance, reconnaissance and targeting systems. In the battlefield context, rapid deployment, self-organization, fault tolerance security of the network should be required. The sensor devices or nodes should provide following services:

- Monitoring friendly forces, equipment and ammunition
- Battlefield surveillance
- Reconnaissance of opposing forces
- Targeting
- Battle damage assessment
- Nuclear, biological and chemical attack detection reconnaissance

13.2.3 Disaster Prevention

The wireless sensor networks are operated for hazardous workspaces like underground mining, steelworks, and refineries. Most of these places entail a high risk by nature. Wireless sensor networks can be deployed in underground mining for surveillance of deteriorating grounds, toxic gases, and unstable grounds. Wireless sensor network periodically gather information for the next course of control action for preventions.

13.2.4 Agricultural Management

The wireless sensor networks can be successfully employed to support resource-poor farmers in developing countries. The resources and technologies available in developing countries towards the usage of wireless sensor network satisfy a growing population's demands for food and other agricultural commodities. The sensor network design support system for resource for the poor farmers that use the wireless sensor network technology for the improvement of farming strategies in the face of highly variable conditions. The use of this application carries out the field study, focuses on water conservation measures and the prediction of crop water requirements for deficit irrigation. Here, the reliable and detailed information gathered by the sensor network about soil moisture and other environmental parameters proved to be the source of these achievements.

13.2.5 Structure Health Monitoring

Wireless Sensor network widely accepted goals of a structure health monitoring system include detecting damage, localizing damage, estimating the extent of the damage and predicting the residual life of the structure. In many old and derelict infrastructures, bridges and railroads, historical buildings such as churches, castles, and monasteries are obliged to be preserved for future generations. Here, seismic and pressure sensors can be deployed to detect and localize stress fractures.

A precise knowledge of stress fractures can be applied for predictive maintenance and for issuing timely warnings to users. The latest approach in this field is promising, because it has many advantages: low deployment and maintenance cost, deployment flexibility, large physical coverage, high special resolution, etc.

13.2.6 Habitat Monitoring

Habitat monitoring is an important mean to better study the behavior of animals with regard to breeding, movement, foraging, etc. This represents a class of sensor network applications with enormous potential benefits for scientific communities and society as a whole. The huge extent of the areas to be covered and the impact of human presence in monitoring plants make the deployment of sensors modules an interesting option. This can lead to distorted results by changing behavioral patterns or distributions or even reduce sensitive populations by increasing stress factors.

For example like the Great Duck Island, a deployment of 32 nodes network to monitor the behavior of storm petrel. To this end the sensors were grouped into sensor patches, and transmit sensor readings to a gateway, which is in turn responsible for forwarding the data from the sensor patch to a remote base station through a local transmit network. The base station, then provides data logging and replicates the data every 15 minutes to a database in Berkeley, California, over a satellite link.

In this way potential relation to typical challenges of developing countries like de-forestation, mono-cropping and new human settlement as well as the global warming problem could be identified. Wireless sensor networks can help to advance in this research area whereby also the whole society can profit by preserving rare species in biological reserves.

13.2.7 Health Applications

Sensor networks are also widely used in health care area. In some modern hospital sensor networks are constructed to monitor patient physiological data, to control the drug administration track and monitor patients and doctors and inside a hospital. In spring 2004 some hospital in Taiwan even use RFID basic of above named applications to get the situation at first hand.

Long-term nursing home: this application is focus on nursing of old people. The room consists of cameras, pressure sensors, orientation sensors and sensors for detection of muscle activity construct a complex network. They support fall detection, unconsciousness detection, vital sign monitoring and dietary/exercise monitoring. These applications reduce personnel cost and rapid the reaction of emergence situation.

13.2.8 Home Application

Along with developing commercial application of sensor network it is no so hard to image that Home application will step into our normal life in the future. Many concepts are already designed by researcher and architects.

In the Fig. 13.2 it shows three different rooms are monitored and controlled by sensor network. At the front door the sensor detects while opening the door, then it will send signal to the electric kettle to boil some water and the air condition to be turned on. The light on the table and is automatically on because the pressure sensor under the cushion has detected weight. The TV gets on. Sensors in the room will be detecting the environment. The air condition will turn to sleep mode until all the sensors get the right temperature.

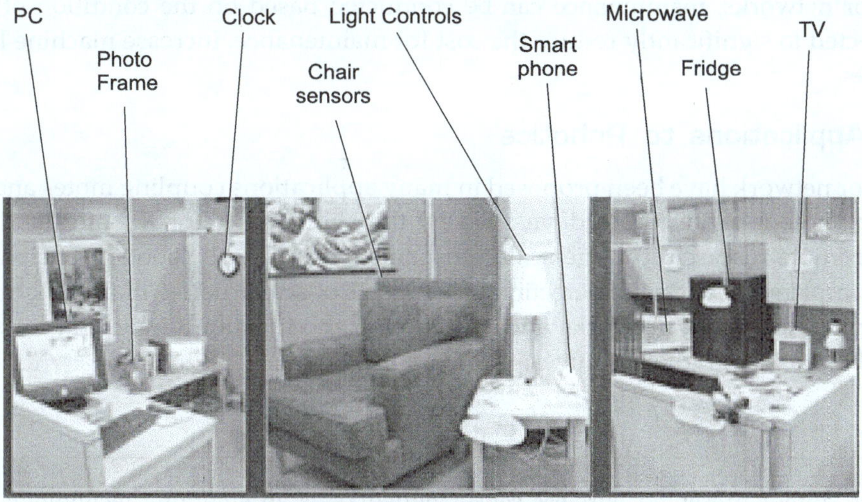

Fig. 13.2 Intel digital homing

The light on the corridor, in the washing room and balcony are all installed with sensor and they can be turned on or turn out automatically. Even, the widows are also attached with vibratory sensors connected to police to against thief.

13.2.9 Industrial Process Control

In industry, WSNs can be used to monitor manufacturing processes or the condition of manufacturing equipment.

Fig. 13.3 An oil pipeline

As shown in Fig. 13.3 wireless sensor networks can be deployed for the flow and checking the chemical plant composition and operations. For example, wireless sensors can be instrumented to production and assembly lines to monitor and control production processes. Chemical plants or oil refiners can use sensors to monitor the condition of their miles of pipelines. Tiny sensors can be embedded into the regions of a machine that are inaccessible by humans to monitor the condition of the machine and alert for any failure.

With sensor networks, maintenance can be conducted based on the condition of equipment, which is expected to significantly reduce the cost for maintenance, increase machine lifetime, and even save lives.

13.2.10 Applications to Robotics

Wireless sensor network have been proposed in many applications coupling motes and robots. For example, Robomote is a tiny robot developed by the USC Center for Robotics and Embedded Systems to promote research in large-scale sensor network, where robots participate. Applications already implemented are the detection of level sets of scalar fields (like isothermal or isobar curves) using mobile sensor networks and imitation of the function of bacteria for seeking and discovering dissipative, gradient sources.

SUMMARY

In this chapter, author presented wireless sensor networks as an emerging technology that has the potential of aiding developing countries to carefully utilize scarce resources, to protect and maintain infrastructures, and to prevent undesirable occurrences. Subsequent, it overview an anatomy and benefits of these networks, we proposed a series of application areas where sensor networks could be most helpful. Chapter covers wireless sensor networks considered for monitoring applications environment, infrastructures, and habitats, agricultural management, and disaster prevention.

Only with collaborative effort the new technology at hand can be leveraged in a sustainable manner. Hence lots of research and development work need to carried out to become complete atomization using wireless sensor network.

QUESTIONS

1. What are the applications of wireless sensor networks?
2. Brief out some examples of real time in the WSN.
3. What are the uses of wireless sensor networks?
4. What are the uses of wirleless sensor networks in the health and habitat monitoring?

BIBLIOGRAPHY

- Akyildiz, I.F., W. Su, Y. Sankarasubramaniam, E. Cayirci, "A Survey on Sensor Networks",IEEE Communications Magazine, August, 102-114(2002).
- Container and Truck Trailer Security Project, Cambridge Systematics, Inc. Parsons Brinckerhoff Quade and Douglas, Inc.
- J. Burrel, T. Brooke and R. Beckwith, "Vineyard Computing: Sensor Networks in Agriculture Production", Pervasive Computing, IEEE Volume 3, Issue 1, Jan.-March 2004 Page(s):38 - 45.

- I.A. Essa, Ubiquitous sensing for smart and aware environments, IEEE Personal Communications (Oct. 2000) 47-49.
- M. Srivastava, R. Muntz, M. Potkonjak, "Smart kindergarten: sensor-based wireless networks for smart developmental problem-solving environments", Proceedings of the 7th annual international conference on Mobile computing and networking.
- Polly Huang, "Sensor Networks Solutions to Real Life Problems".
- A Survey of Applications of Wireless Sensors and Wireless Sensor Networks; Th. Arampatzis, J. Lygeros, Senior Member, IEEE, and S. Manesis, Member, IEEE; Proceedings of the 13th Mediterranean Conference on Control and Automation Limassol, Cyprus, June 27-29, 2005.
- Chee-Yee Chong; Kumar, S.P., "Sensor networks: Evolution, opportunities, and challenges,"Proc IEEE, August 2003
- M. Beigl, A. Krohn, T. Riedel, T. Zimmer, C. Decker, M. Isomura The uPart Experience: Building a Wireless Sensor Network, IEEE/ACM, Conference on Information Processing in Sensor Networks (IPSN), 2006.
- Wireless Sensor Networks: Concepts and Applications, Silicon Valley Technical Institute 1762 Technology Drive San Jose, CA.
- Design and Evaluation of a Wireless Sensor Network Based Aircraft Strength Testing System, Jian Wu, Shenfang Yuan, Genyuan Zhou, Sai Ji, Zilong Wang and Yang Wang.
- Literature Survey on Wireless Sensor Networks, Pavlos Papageorgiou, July 16, 2003.
- I.F. Akyildiz, Weilian Su, Y. Sankarasubramaniam, and E. Cayirci. A survey on sensor networks, IEEE Communications Magazine, Aug. 2002.
- D. Estrin, R. Govindan, J. Heidemann, and S. Kumar. Next century challenges: Scalable Coordination in Sensor Networks, In ACM MobiCom, 1999.

INDEX